Advanced Science

Electrical Installation Series – Advanced Course

C. Duncan

Edited by Chris Cox

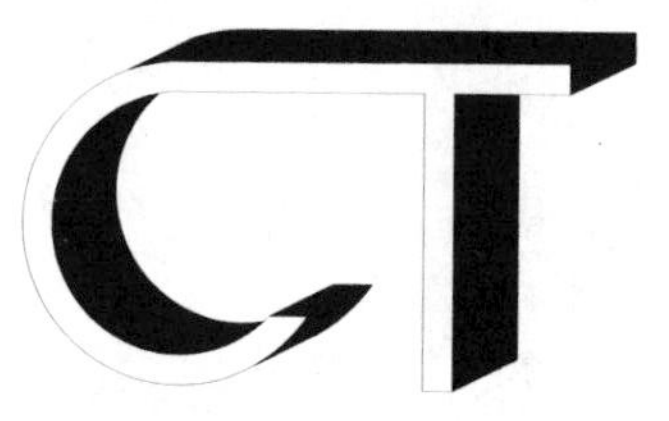

THOMSON LEARNING

Australia · Canada · Mexico · Singapore · Spain · United Kingdom · United States

Advanced Science

For more information, contact Thomson Learning, Berkshire House, 168–173 High Holborn, London, WC1V 7AA or visit us on the World Wide Web at: http://www.thomsonlearning.co.uk

---

British Library Cataloguing-in-Publication Data
A catalogue record for this book is available from the British Library

---

ISBN 1-86152-730-6

First published 2001 by Thomson Learning

Printed in Croatia by Zrinski d.d.

About this book

"Advanced Science" is one of a series of books published by Thomson Learning related to Electrical Installation Work. The series has been designed for those who are studying for courses such as the City and Guilds 2360 - Electrical Installation Work - Course C. It is also suitable as a "stand alone" module for short courses and for use on some BTEC modules.
A complete list of titles in the series is given below.

Electrical Installation Series

Foundation Course

Starting Work
Procedures
Basic Science and Electronics

Supplementary title:
Practical Requirements and Exercises

Intermediate Course

The Importance of Quality
Stage 1 Design
Intermediate Science and Theory

Supplementary title:
Practical Tasks

Advanced Course

Advanced Science
Stage 2 Design
Electrical Machines
Lighting Systems
Supplying Installations

Study guide

This studybook has been written to enable you to study the theory related to electrical and electronic science either in a classroom or in a private study situation. To ensure that you gain the maximum benefit from the material you will find prompts all the way through that are designed to keep you involved with the subject. If you are studying by yourself the following points may help you.

- ☞ Work out when, and for how long, you can study each week. Complete the table below and from this produce a programme so that you will know approximately when you should complete each chapter. Your tutor may be able to help you with this. It may be necessary to reassess this timetable from time to time according to your situation.
- ☞ Try not to take on too much studying at a time. Limit yourself to between 1 hour and 2 hours and finish with a Try this or the short answer questions (SAQ) at the end of the chapter. When you resume your study go over this same piece of work before you start a new topic.
- ☞ You will find answers to the questions at the back of the book but before you look at the answers check that you have read and understood the question and written the answer you intended.
- ☞ Questions are included at the end of each chapter and at the end of the book so that you can assess your progress.
- ☞ Try this activities are included and you may need to ask colleagues at work or your tutor at college questions about practical aspects of the subject. These are all important and will aid your understanding of the subject.
- ☞ It will be helpful to have available for reference a current copy of BS 7671:1992. At the time of writing this incorporates Amendment No.1, 1994 (AMD8536), Amendment No. 2, 1997 (AMD 9781) and Amendment No. 3, 2000 (AMD 10983).
- ☞ Your safety is of paramount importance. You are expected to adhere at all times to current regulations, recommendations and guidelines for health and safety.

Study times					
	a.m. from	to	p.m. from	to	Total
Monday					
Tuesday					
Wednesday					
Thursday					
Friday					
Saturday					
Sunday					

Programme	Date to be achieved by
Chapter 1	
Chapter 2	
Chapter 3	
Chapter 4	
Chapter 5	
Chapter 6	
Chapter 7	
End test	

Contents

Introduction

Although this studybook covers theory related to electrical and electronic science, there are a number of laboratory/workshop activities that need to be carried out to complete the Course C syllabus. These exercises need to be covered in a recognised centre and should be carried out under the advice of a tutor.

Whenever practical work is carried out full compliance with the Health and Safety at Work Etc. Act 1974 and the Electricity at Work Regulations 1989 should be observed.

Laboratory/Workshop Activities related to this studybook as identified in the City and Guilds syllabus:

- Murray loop test for cable fault location
- Charge and discharge curves (capacitor)
- Current in a neutral (3 phase unbalanced load) and the effect of a broken neutral
- Verification of Kirchhoff's laws
- R L C series and parallel circuits
- Resonance (i.e. variable frequency and variable reactance)
- Efficiency and regulation of a transformer
- 3 phase power factor improvement
- Use of CTs and VTs
- Two-wattmeter method (3 phase power)
- Use of Wheatstone bridge
- Use of cathode ray oscilloscope to determine the amplitude and frequency of alternating voltages
- Introduction to the use of manufacturers' data to identify connections to semiconductor devices
- Construction of a simple Zener diode stabilising circuit
- Investigation of the operation of a simple amplifier to determine amplifier gain
- Production of a simple astable multi vibrator to demonstrate the relationship between capacitance, and resistance and the markspace ratio
- Construction and testing of basic logic circuits
- Use of test instruments to perform simple tests on semiconductor devices

1

Electromagnetism

At the beginning of all the other chapters in this book you will be asked to complete a revision exercise based on the previous chapter.

To start you off we will use this opportunity to remember some facts about multiples and submultiples of basic units. For ease of calculation we use "powers of ten".

For example:

mega (M) = 10^6
kilo (k) = 10^3
milli (m) = 10^{-3}
micro (μ) = 10^{-6}

Therefore a 10 μF capacitor is written as

10×10^{-6} Farads

Similarly a 250 mH inductor is written as

250×10^{-3} Henries and so on.

If you are using a scientific calculator with an EXP or EE button try using that.

For example:

4×10^{-6}
4 EXP –6 (you don't have to put in the 10).

If you are still having problems, talk to your tutor.

On completion of this chapter you should be able to:

- identify the basic rules of electromagnetism
- describe a magnetic circuit
- calculate the effectiveness of a magnetic circuit
- describe the relationship between magnetism, current flow and mechanical movement
- describe the effects of self inductance

The effects of magnetism have been known to mankind since early navigators discovered that pieces of a certain type of mineral, when suspended from a thread, would always point in the same geographical direction. "Lodestone" was used as a simple but effective magnetic compass which possessed other seemingly magical properties, such as the power to attract small iron objects and to move other lodestones without actually touching them. In our present state of apparent enlightenment, even small children will explain these mysteries to their elders as the alignment of the magnetic field of the compass with that of the planet Earth, or the forces of attraction and repulsion between magnetic poles.

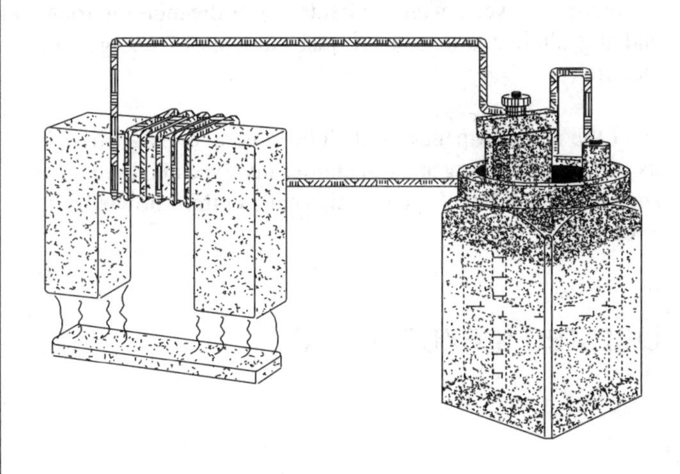

Figure 1.1

Most people who are familiar with magnets are aware of the fact that the magnet has two poles. Either pole will attract unmagnetised pieces of iron or steel but if one magnet is brought close to another, opposing poles will attract each other with considerable force whereas like poles will exert a force which prevents them from coming together.

Every magnet has around itself a space which is charged with energy and is called a "magnetic field". If a piece of unmagnetised iron is brought into this field then it becomes magnetised but in most cases when it is taken out of the field it loses its magnetism again. This is not always the case however, as we are all familiar with some hard steel objects, such as screwdriver blades, which remain magnetised and prove to be quite useful when retrieving small nuts, screws or washers from the innards of equipment where they should never have been dropped in the first place. From this, it can be deduced that whether a magnet is to be permanent or not depends on the nature and composition of the metal. A soft ferrous object such as a nail will quickly lose any magnetism it may have acquired. Steel objects such as tools will tend to retain their magnetism and may have to be demagnetised if this proves to be a nuisance.

There was undoubtedly a suspicion that a link existed between electricity and magnetism but it was not proved until the Danish physicist Hans Oersted in 1819, showed that a current-carrying conductor had a magnetic field around it which ceased when the current stopped and changed direction when the current was reversed. This proved to be a major breakthrough in technology for shortly afterwards in 1831, Michael Faraday discovered that, not only could magnetism be produced by electricity, but electricity could be produced from magnetism.

Today's electrician is well aware of the connection between electricity and magnetism as much of our work involves getting supplies to coils and windings of all descriptions.

The magnetic force of attraction causes a relay to close. The interaction between magnetic fields causes the motor to rotate and the alternator to convert vast amounts of energy into electricity.

Over the next few pages we shall be looking at magnetism in its own right, looking at the quantities involved and their units and finding out how the electrician can put this knowledge to good use.

Oersted's experiment

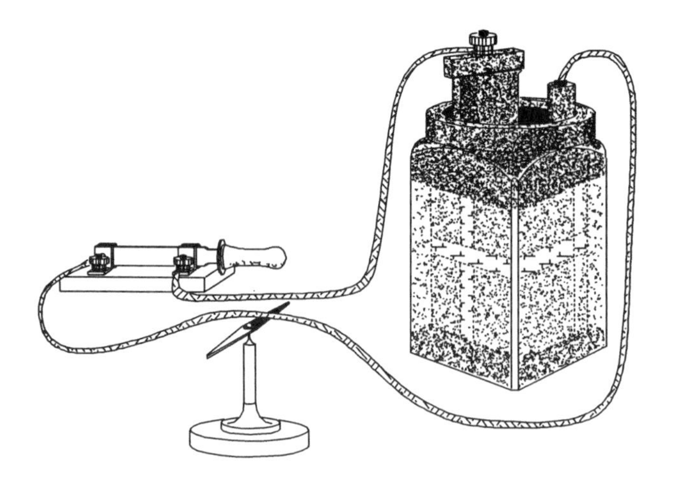

Figure 1.2

Try this

Connect a load to a d.c. supply, for example a lamp to a battery. Pass a compass needle slowly across the two wires in turn and you should see the needle reverse direction as it goes from one wire to the other.

At least it did when the author did it!

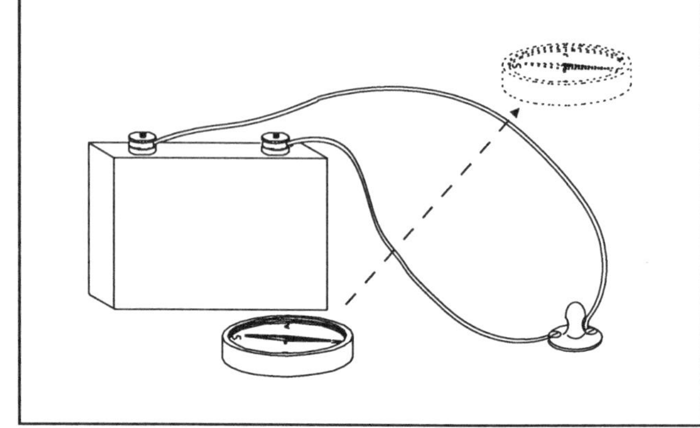

Magnetic flux

Within the area surrounding a magnet, i.e. the magnetic field, the magnetic force is grouped into lines. There is nothing physical about these lines that you will ever see or touch but their presence is real enough and you can trace their outline by sprinkling iron filings on a sheet of paper and placing this over a magnet. A gentle tap on the paper will help the iron filings to find the direction of the flux.

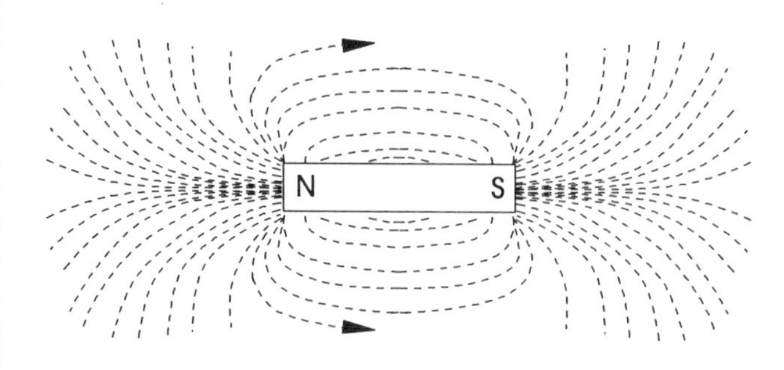

Figure 1.3 *Magnetic field around a permanent magnet.*

The total amount of magnetism in the field is known as the **FLUX**.

It is given the symbol ϕ and is measured in Webers (abbr. Wb). This is phi, ϕ, the Greek letter 'f'.

Remember

There are four important facts about lines of magnetic flux:

1. **The direction of the line of magnetic flux at any point in its path is the same as would be shown by a compass needle at that point.**
2. **Lines of magnetic flux outside the magnet start at a north pole and finish at a south pole.**
3. **The lines never cross but always run roughly parallel to each other.**
4. **Lines of magnetic flux which are running parallel and in the same direction repel each other.**

NOTE

The North pole of a magnet is really a "north-seeking" pole which is attracted and not repelled by the Earth's magnetic North pole.

Try this

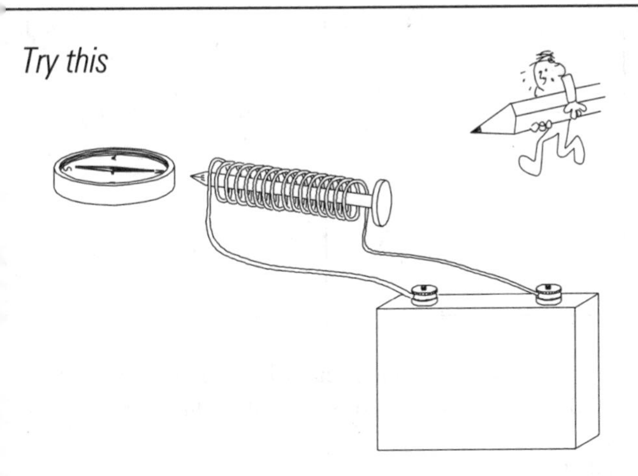

Take a piece of insulated wire, wrap it round a nail, connect it to a battery and test the ends of the nail with a compass needle.

Reverse the battery connections and test again.

It is now obvious that, with the right equipment, an electric current can produce a magnetic flux, not only in the nail but in the space surrounding it as well.

A small current will produce a small amount of magnetic flux and a larger current, a proportionally greater amount.

Of course there are a lot of variables in between but basically this is true.

NOTE:
THE MAGNETIC FLUX IS PROPORTIONAL TO THE CURRENT PRODUCING IT.

Magnetic flux density

The effects of the flux are most pronounced where the lines are grouped closely together such as at the ends of the magnet, near the poles. As you move away from the magnet the field becomes weaker and the flux produced by the magnet becomes merged with the Earth's magnetic flux. The total amount of flux is not going to have much effect if it is dispersed over a wide area but concentrated in a small space, such as in the metal core or at the ends, it may have considerable influence.

As this is an important feature of the magnetic field it is regarded as a definable quantity and given its own unit.

A total flux of **one Weber** concentrated into an area of **one square metre** is said to have flux density of **ONE TESLA**.

The symbol for flux density is B

and the abbreviation for the Tesla is T

This gives us the equation:

$$B = \frac{\phi}{a}$$

Where				
	B	=	flux density in Teslas	T
	ϕ	=	flux in Webers	Wb
	a	=	area in square metres	m^2

Example

If a magnetic flux of 0.45 Wb is concentrated in an area of 0.3 m^2 the flux density in this part of the field is:

$$B = \frac{\phi}{a} = \frac{0.45}{0.3} = 1.5\text{T}$$

Now follow through this example:

How much flux would be required to produce a flux density of 2.2 T at a magnet face measuring 4 cm × 6 cm?

$$a = 0.06 \times 0.04 = 0.0024 \text{ m}^2$$

$$B = \frac{\phi}{a}$$

which transposed becomes

$$\phi = B \times a = 2.2 \times 0.0024 = 0.00528 \text{ Wb}$$

From this you will realise how important it is to stick to your units and convert these back into basic units for calculation purposes. Multiples and sub-multiples can always be used in your final answer.

For example, 0.00528 Wb = 5.28 mWb (milliwebers)

That's enough about flux density for the time being, let's take a look at something else.

Try this

1. A magnet has a flux of 6.8 mWb and a face of 5 cm by 4.5 cm. Calculate the flux density of the pole.

2. Calculate the flux required to produce a flux density of 0.5 T at a magnetic face of 2 cm by 2.5 cm.

3. A magnetic pole face produces a flux density of 4.2 T and has a flux of 4 mWb. If one side of the pole face is 2 cm calculate the other.

The magnetic circuit

In much the same way as we regard the conductors of an electric circuit as a means of conveying an electric current from one part of the circuit to another, we look to the core of a magnetic circuit as a means of carrying magnetic flux.

If the flux can be compared to an electric current then the other two Ohm's Law circuit quantities will have their equivalents.

ELECTRIC	MAGNETIC
e.m.f. E, V or U volts (electromotive force)	m.m.f. F amperes* (magnetomotive force)
Current I amperes	Flux ϕ Wb
Resistance R ohms	Reluctance S
$I = \frac{U}{R}$ (A)	$\phi = \frac{F}{S}$ (Wb)
R, U, I	S, F, ϕ

The m.m.f. of a magnetic circuit will depend on the current and the number of turns of wire in the coil.

***NOTE**
As far as the magnetic circuit is concerned it does not matter whether this is a current of one quarter of an ampere going round the core forty times, or ten amps going round once; so the m.m.f. is simply quoted in amperes. But remember, in practice you must **always** multiply the actual current in the coil by the number of turns.

Example
A coil of 40 turns passing a current of 0.25 A produces an m.m.f. of:

$$F = N \times I$$
$$= 40 \times 0.25$$
$$= 10 \text{ A}$$

Now let's see how this would work out.

A coil of 500 turns is wound around a ring core which has a reluctance of 0.4. What is the flux in the core when the coil is carrying a current of 0.5 mA?

$$\phi = \frac{F}{S*}$$

$$= \frac{500 \times 0.0005}{0.4}$$

$$= 0.625 \text{ Wb or } 625 \text{ mWb}$$

*There are no "real units" for reluctance so the usual practice is to say "S = a number", otherwise it becomes more confusing than it needs to be.

Example

A coil of 300 turns is wound on a core having a reluctance of 1.8. How much current must the coil carry to produce a flux of 750 mWb in the core ?

$$\phi = \frac{F}{S}$$

which transposed becomes

$$F = \phi \times S$$
$$= 0.75 \times 1.8$$
$$= 1.35 \text{ A}$$

But $F=NI$; therefore the current in the coil will be divided by the number of turns.

$$I = \frac{1.35}{300}$$

$$= 0.0045 \text{ A} \qquad \text{or } 4.5 \text{ mA}$$

Try this

1. A coil with 750 turns is wound on a core with a reluctance of 1.2. What flux is produced when a current of 0.6 mA is flowing?

2. A coil with 810 turns has a resistance of 200 kΩ and is wound on a core with a reluctance of 3.8. What flux is produced when the coil is connected to a 24 V supply?

Resistivity and permeability

Whilst considering the comparison between electrical and magnetic quantities we cannot overlook these two:

Resistivity tells you how bad a conductor can be:
i.e. high resistivity – poor conductor material,

Permeability is an indication of how well a magnetic material will perform:
i.e. high permeability - easily magnetised.

To determine the resistance of a conductor you multiply the resistivity by the length and divide by the cross-sectional area.

$$R = \frac{\rho \times \text{length}}{\text{area}} \quad (\rho \text{ RHO})$$

To determine the reluctance of a magnetic path you must divide the length by the product of the permeability and the cross-sectional area.

$$S = \frac{\text{length}}{\mu \times \text{area}}$$

Where μ (pronounced "mu") is the symbol for permeability.

Example

A piece of iron is 0.15m long and has a cross-sectional area of 0.001 m^2. In this example the permeability is taken as

$2\pi \times 10^{-4}$.

This gives a value for S of

$$S = \frac{0.15}{2\pi \times 10^{-4} \times 0.001}$$

$$= 238732$$

Permeability is actually made up of two parts:

1. μ_0 which is the permeability of any non-magnetic medium such as air, wood, carrot or cucumber. This is a constant i.e. it is always the same value.

   $\mu_0 = 4\mu \times 10^{-7}$

   in other words 1.256×10^{-6}

(This figure is nearly always quoted in examination questions on magnetism so there is no need for you to memorise it.)

2. The second part μ_r is the relative permeability of the magnetic material in question. This varies from 1 for free space or non-magnetic media, through 370 for mild steel up to values in excess of 10000 for some specialised magnetic alloys.

To calculate the absolute permeability of the magnetic material, these two values are multiplied together giving:

$$\mu = \mu_0 \times \mu_r$$

In the above example, if we used a relative permeability of 500 this would have given us:

$$\mu = \mu_0 \times \mu_r$$

$$= 4\pi \times 10^{-7} \times 500$$

$$= 6.28 \times 10^{-4}$$

which was the figure used.

So if this is the equivalent resistance of the magnetic circuit, it must be possible to substitute the other values for the equivalents of current and voltage.

Let's widen the picture a little and find out some more about this particular magnetic circuit.

Firstly: it is in the form of a closed ring, so the length of the circuit is the mean circumference.

Secondly: it is wound with a coil of 250 turns of wire.

Thirdly: the flux density in the core is to be 1.5 Teslas.

What we really want to know is: How much current must we put into the coil to get this result?

Reluctance – is equivalent to resistance –

$S = 238732$ (previously calculated)

Flux – is equivalent to current

$\phi = B \times a$

$= 1.5 \times 0.001$

$= 1.5 \times 10^{-3}$ Wb

$\phi = 1.5 \times 10^{-3}$

Magnetomotive force – is equivalent to voltage

Since $U = I \times R$

m.m.f. $= \phi \times S$

$= 1.5 \times 10^{-3} \times 238732$

$= 358$ ampere turns

As there are 250 turns of wire in the coil, this means that the current will have to be:

$$= \frac{358}{250}$$

$$= 1.432 \text{ A}$$

Composite magnetic circuits

In the same way as some electrical circuits are made up of several components connected in series and/or parallel, magnetic circuits can be similarly constructed.

Try this

A magnetic pole 0.05 m long and 0.0025 m^2 cross-sectional area is made from steel with a relative permeability of 600. If the coil is wound with 450 turns and the flux density is to be 2.25 Teslas, calculate the current that the coil will take.

The magnetisation curve

This is also known as the *B*/*H* curve and shows the relationship between flux density and magnetising force for a specific type of magnetic material.

The quantity H, i.e. the magnetising force, is the ampere turns per metre length of the magnetic path.

Taking the previous example:

The current was 1.43 A
The number of turns 250
The circuit length 0.15 m

Therefore the magnetising force is

$$\frac{143 \times 250}{0.15} = 2383$$

H = 2383 Ampere turns per metre

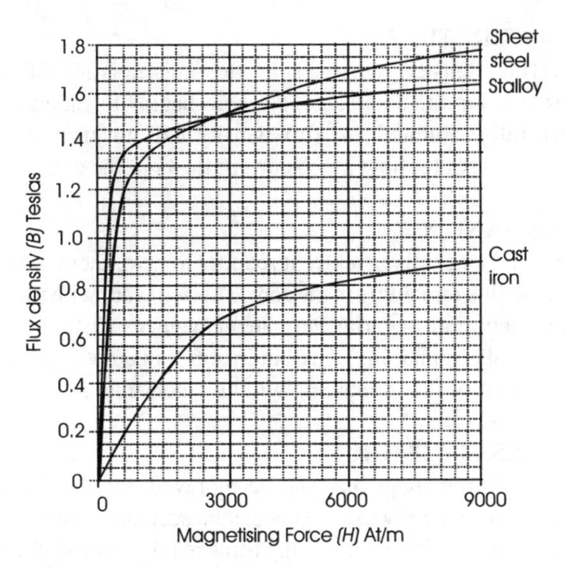

Figure 1.4 *A typical B/H curve*

You will see that the tops of the curves flatten out. This is because the core becomes "saturated" and cannot carry any more magnetic flux even though the current in the coil carries on increasing. Some materials saturate before others for example cast iron, and some are more easily magnetised (more permeable) for example stalloy.

These features can be seen on the *B*/*H* curve showing comparative examples (Figure 1.4).

Hysteresis

As the current is increased the flux density increases. As the current is decreased, the flux density decreases but it does not follow the same path. On the downward path the flux density is higher than on the way up.

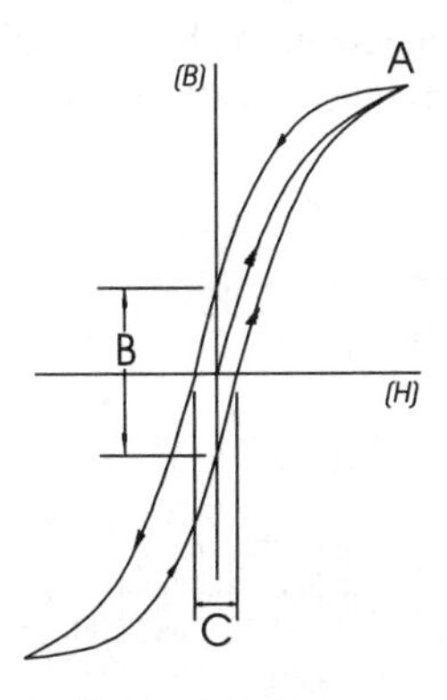

Figure 1.5 *Typical hysteresis loop*

When a magnet is connected to an a.c. source the flux in the core is increased then decreased and reversed and brought back to zero again in the course of each cycle of the supply. This gives the characteristic form known as the hysteresis loop as shown here. There are several features of the loop which can convey information about the properties of the magnetic material.

A. SATURATION
The flux density increases as the magnetising force is increased but at some stage, the core begins to saturate and when fully saturated, the flux density will not increase any further, no matter how much current flows in the coil.

B. REMANENCE
When the magnetising current reduces to zero, there will still be some flux remaining in the core. This will vary with the type of magnetic core material. Permanent magnets will have a high "remanent" (remaining) flux but in softer magnetic materials the amount of flux remaining will be much smaller.

C. COERCIVE FORCE
Before the remanent flux can be removed, the magnetising current has to be reversed. This can be seen on the hysteresis loop as the negative magnetising force required before the flux density passes through zero. The coercive force varies enormously between permanent magnetic materials and those, for example, used in transformer cores.

Airgaps

Airgaps are a necessary part of many magnetic circuits. For example:

- There has to be an airgap between the fixed and rotating parts of an electric motor to allow rotation to take place.
- The two halves of a relay assembly have a considerable airgap when the relay is open but none when the relay closes.

Airgaps in a magnetic circuit are similar in effect to resistors in an electrical circuit. This is due to the fact that some magnetic materials are up to ten thousand times better than air at carrying magnetic flux.

An airgap can be introduced to allow variation in the reluctance of the magnetic path. A variable airgap can be a useful device when controlling the inductance of an iron-cored coil.

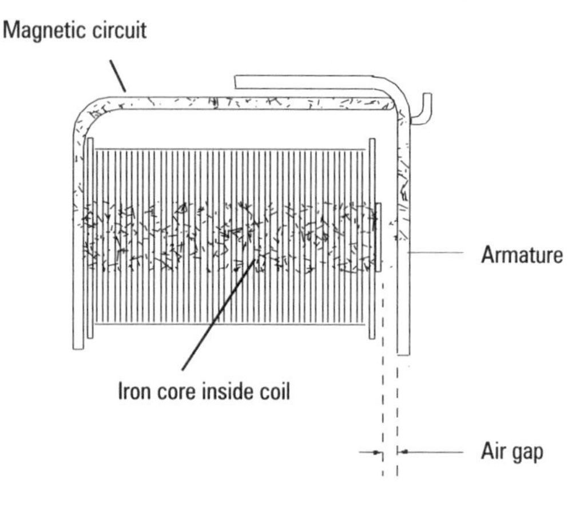

Figure 1.6 *Electro-magnetic relay*

Force on a conductor

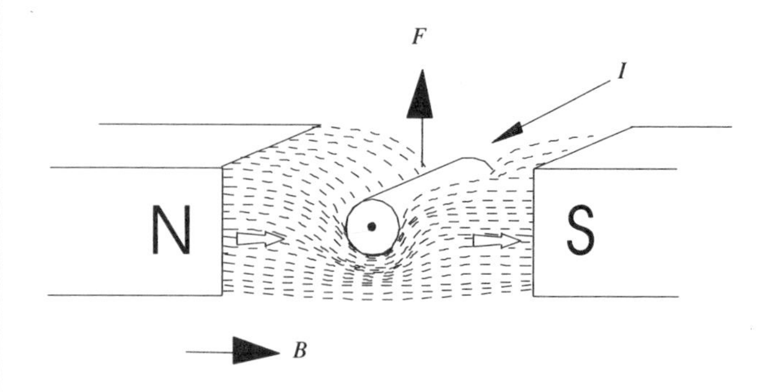

Figure 1.7 *The force produced on a current carrying conductor in a magnetic field*

A straight, current-carrying conductor situated in the gap between the poles of a magnet is subjected to a force in Newtons equivalent to the product of its length, the flux density and the current in the conductor.

$$F = B\,I\,L$$

Where			
	F	= Force in Newtons	N
	B	= Flux density in Tesla	T
	I	= Current in Amperes	A
	L	= Length in metres	m

The direction of the force can be determined by Fleming's Left-Hand Rule

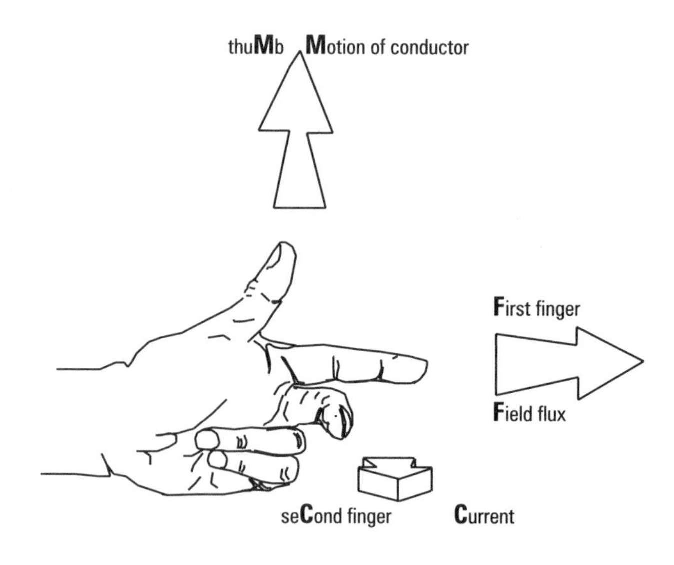

Figure 1.8 *Fleming's left-hand rule*

Induced e.m.f.

A straight conductor passing between the poles of a magnet has an e.m.f. induced in it which is equivalent to the product of the flux density, the length of conductor in the field and its velocity.

$$E = B\,L\,v$$

Where E = Induced e.m.f. in volts V
B = Flux density in Tesla T
L = Length in metres m
v = Velocity in metres/sec m/s

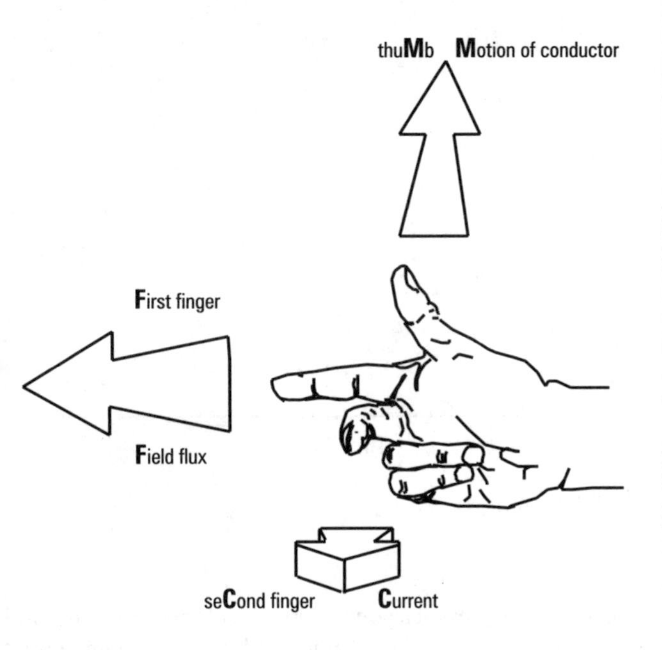

Figure 1.9 Fleming's right-hand rule

The direction of the induced e.m.f. can be determined by Fleming's Right Hand Rule.

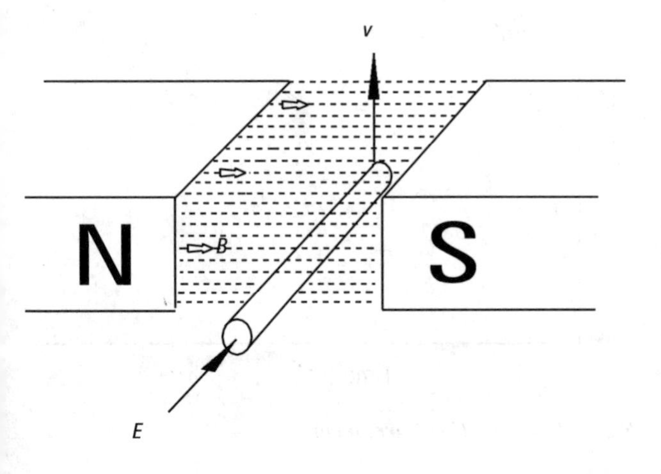

Figure 1.10 Induced e.m.f.

Self inductance

As discussed earlier, a current-carrying conductor possesses its own magnetic field. When this conductor is wound into a coil, the field produced has the properties of a magnet. This becomes more pronounced if the coil is wound around an iron core. When the current is first switched on the magnetic field expands outwards, cutting the conductors and producing an induced e.m.f. This has the effect of opposing the e.m.f. producing the rise in flux density.

A German physicist, Heinrich Lenz said the same thing (in German) in about the year 1834 and this became known as Lenz's Law:

"The direction of an induced e.m.f. is always such that it tends to set up a current opposing the motion or the change in flux responsible for producing that e.m.f."

An inductor is said to have an inductance of **one Henry** when a current which is changing at the rate of **one Ampere per second** produces an induced e.m.f. of **one volt.**

The Henry is the unit of Inductance and you will find it used to evaluate the inductive properties of chokes and coils in a variety of applications. One Henry is quite a large value in practical terms and it is usual to find inductors measured in millihenries for everyday applications.

When an inductive circuit is first switched on, the induced e.m.f opposes the increase in current. Instead of immediately rising to its final value, the current rises more gradually, at a rate governed by the inductance of the circuit.

For example, a circuit having an inductance of 250 mH and a resistance of 250 mΩ is connected across a 1 V supply.

The final circuit current is

$$I = \frac{U}{R}$$

$$= \frac{1}{0.25}$$

$$= 4\ \text{A}$$

The only difference between this and a purely resistive circuit is the time taken for the current to reach this value.

The time constant

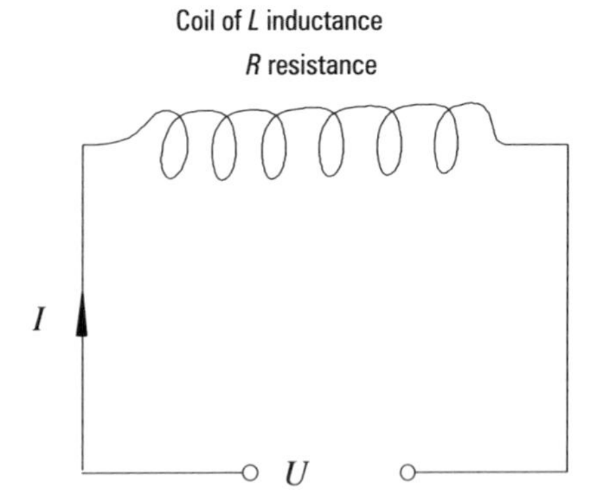

Figure 1.11

The rate at which the current increases can be determined by the expression:

$$\text{Time constant} = \frac{\text{inductance in Henries}}{\text{resistance in Ohms}}$$

Time constant has the symbol τ (Greek letter "TAU") and is measured in seconds (s).

$$\tau = \frac{L}{R}$$

This gives the time it would take for the current to reach its final value if it were to increase at its initial rate.

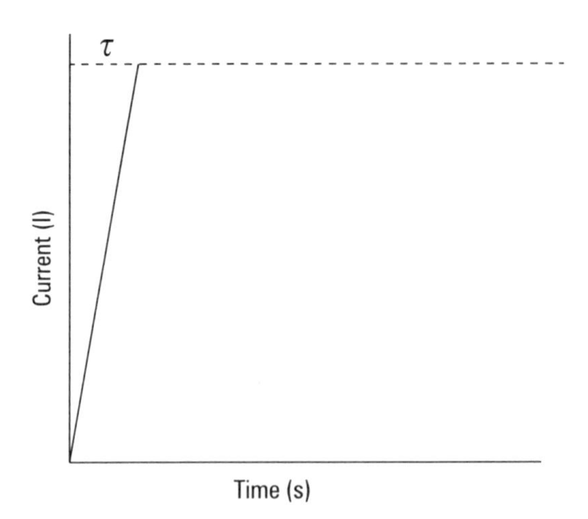

Figure 1.12 *Time constant*

Current growth curve

A good approximation of the actual growth of current can be obtained by the following method.

Draw the first line as before.

Then, at a point "**P**" about one quarter of the way up that line mark off a distance equal to one time constant. Mark vertically upwards until you reach the maximum current line, mark this point "**Q**": Join **P – Q**.

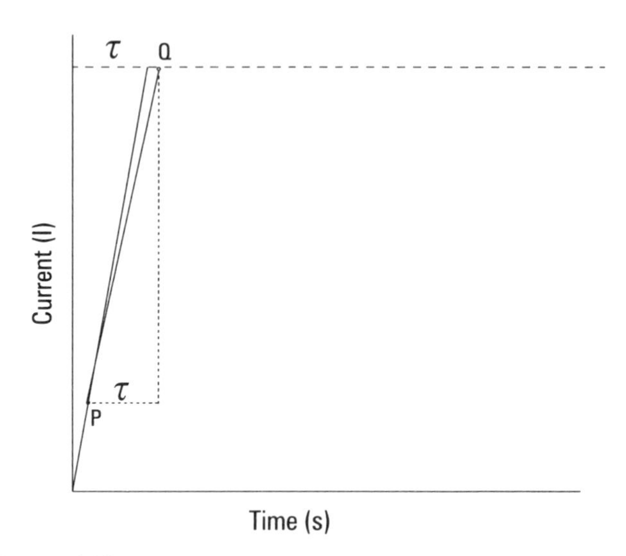

Figure 1.13

Repeat this process at a point on **P – Q** and keep repeating until the current growth curve merges with the maximum current line.

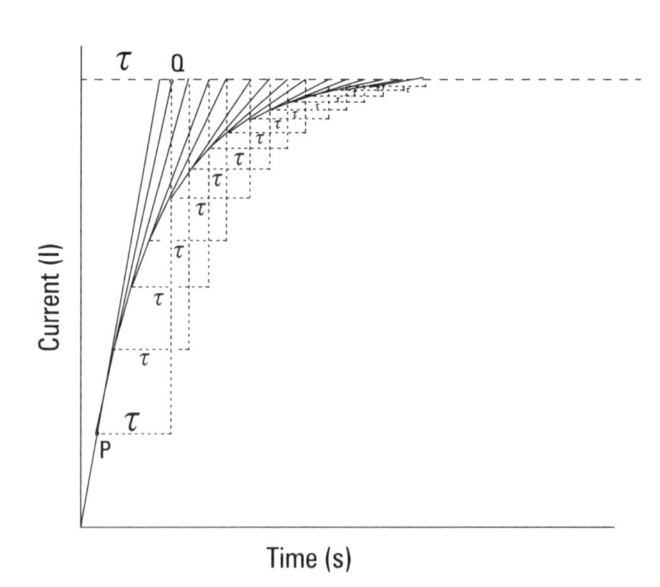

Figure 1.14 *Time/current curve*

Using the current growth curve

This graph can now be used to determine:

(a) the current at any given time

or

(b) the time taken to reach a given current after the circuit has been switched on.

Example

A 100 V d.c. circuit has a resistance of 10 Ωand an inductance of 1.5 H.

Determine

(a) the current 0.19 seconds after switching on and

(b) the time taken for the current to reach 5 A.

Obviously we will have to draw a growth of current graph for this circuit and before we can do this we will have to find out two things about the circuit.

i) the time constant and

ii) the final current

i) time constant

$$= \frac{L}{R} \text{ seconds}$$

$$= \frac{15}{10}$$

$$= 0.15 \text{ s}$$

ii) final current

$$= \frac{U}{R}$$

$$= \frac{100}{10}$$

$$= 10 \text{ A}$$

Time/Current Graph (typically four to six constants long)

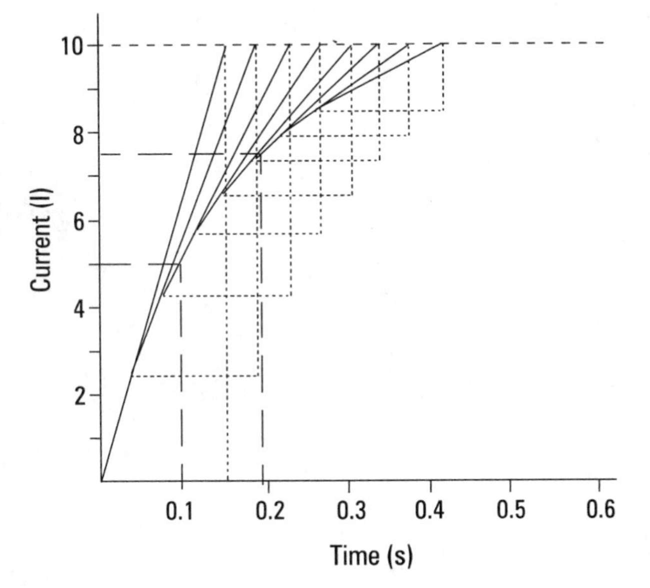

Figure 1.15

Answer

From the graph we can estimate that

(a) the current is 7.6 A 0.19 seconds after switching on

(b) it takes 0.1 seconds for the current to reach 5 Amps.

Energy stored in an inductive circuit

During the time taken for an inductive circuit to reach maximum current after switching on, energy from the electrical input has been transferred to the magnetic field. Whilst the current is flowing, this energy level is maintained and held in storage in the inductor.

The amount of energy is determined by the current and the inductance and can be calculated by:

Energy $W = \frac{1}{2} L I^2$ Joules

Example

A coil of 250 mH carries a d.c. current of 4 A. How much energy is stored in the coil while this current is flowing?

$$W = \frac{1}{2} L I^2 \text{ Joules}$$

$$= 0.125 \times 16$$

$$= 2 \text{ Joules}$$

The presence of this stored energy does not normally cause a problem. Indeed it is the stored energy in an inductor that helps to smooth the output of many power supply units.

The problem arises in the switching of inductive circuits.

When switched off, the inductor releases its energy back into the circuit in the form of a high induced voltage. This is exactly what occurs when the contact breaker points open in a car or motor-bike engine. As the points open, a surge of high voltage is directed to the spark plugs and the resultant spark discharge ignites the fuel mixture.

Imagine what would happen if a voltage of many thousands of volts suddenly appeared amongst a load of delicate electronic components. That's when special precautions have to be taken to suppress this induced voltage before it can do any damage.

Try this

Make a list of applications that switch highly inductive loads.

Protection for inductive circuits

Flywheel diode

A device which is frequently used to suppress induced voltages when switching highly inductive circuits such as motors is the "flywheel diode".

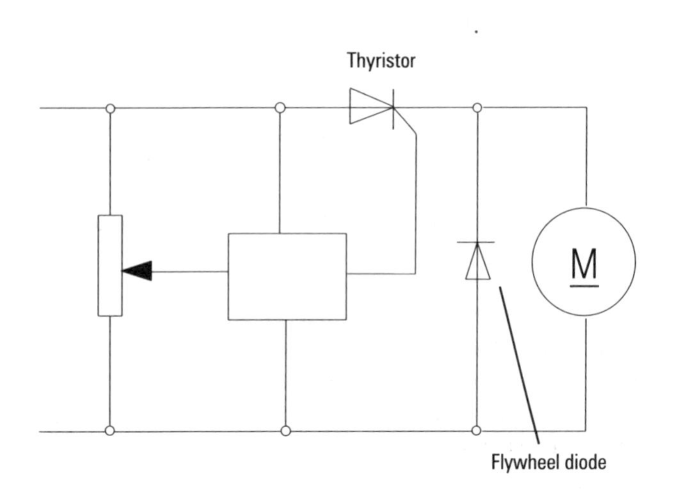

Figure 1.16 Typical circuit using flywheel diode

As the thyristor switches off the load, the induced voltage feeds a current back into the circuit which can switch the thyristor back on when it should be off.

If a flywheel diode is connected across the output of the switching circuit, i.e. in parallel with the inductive load, then this conducts the induced current back into the load until it is completely dissipated and is no longer a problem.

Voltage-dependent resistor

The voltage-dependent resistor (v.d.r.) is a device, the resistance of which varies with the potential difference applied across its terminals. High induced voltages which normally accompany the switching of inductive circuits will produce a lowering of the v.d.r.'s resistance and this will quickly dissipate the energy in the inductor. Once the voltage is restored to normal, the v.d.r. returns to its basic state.

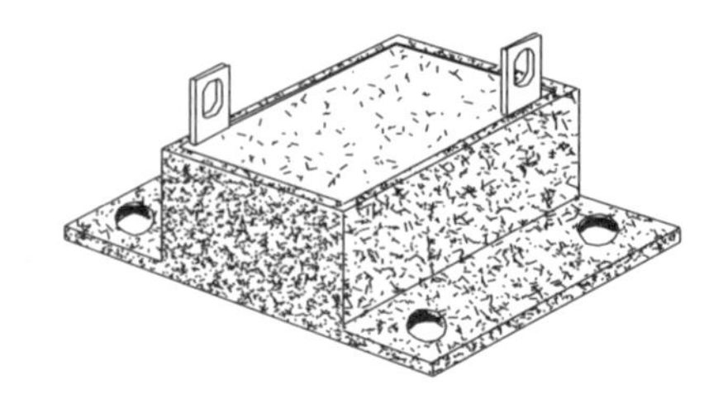

Figure 1.17 High energy

Figure 1.18 Small power

Exercises

1. (a) Draw a typical B/H curve for a sample of magnetic steel and indicate what each part of the curve represents.

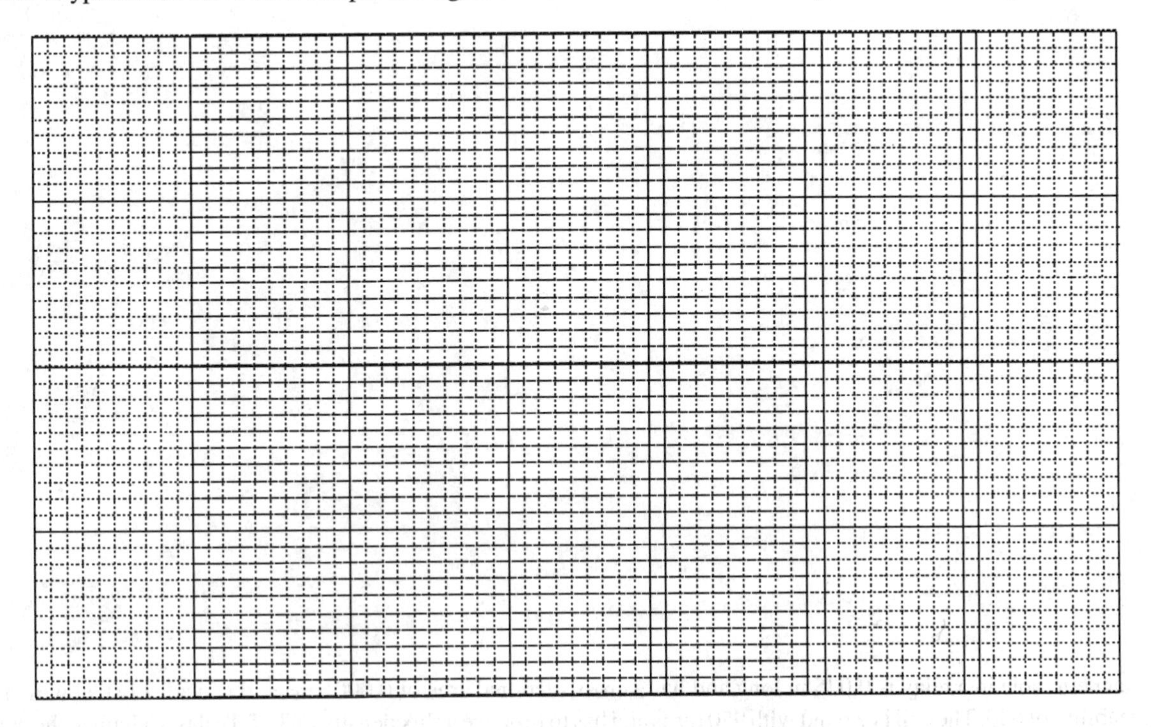

(b) State the units that are used to measure

i) B

ii) H

(c) Explain, with the aid of a diagram, the relationship between a *B*/*H* curve and a hysteresis loop.

(d) Explain the purpose of a "flywheel diode" in an inductive circuit.

2. (a) Explain the term "time constant", as applied to a circuit containing resistance and inductance in series.

(b) A coil with an inductance of 2.5 H and resistance of 25 Ω is connected to a 50 V d.c. supply. Determine

   i) the time constant for this coil
   ii) from a graphically drawn curve, how the current varies with time to 0.5 s after the current is switched on.
   iii) Determine the value of the current 0.2 s after the current is switched on.

3. (a) A magnetic pole in a relay is 0.025 m long and has a cross sectional area of 0.0001 m^2 and is made of steel with a relative permeability of 650. The coil is wound with 950 turns and has to produce a flux density of 3.15 Teslas. Calculate the current the coil will take to make the relay operate.

(b) Explain the general advantages of using relays on d.c. supplies instead of a.c.

(c) Explain why the air gap between the armature and iron core in a relay is kept to a minimum.

2

Electrostatics

Complete the following to remind yourself of some important facts on this subject that you should remember from the previous chapter.

1. If $\phi = 0.85$ and $a = 0.7$ then $B =$

2. What is the symbol used to signify the permeability of any non-magnetic material?

3. In "Fleming's Rule", either hand, the thumb indicates

4. What is the Greek letter τ used to represent?

On completion of this chapter you should be able to:

- describe the basic theory relating to electrostatics
- identify different types of capacitor from its case
- calculate the different effects of connecting capacitors in series and parallel
- calculate the energy stored in given capacitors
- describe the properties of different types of capacitor
- describe the charge and discharge characteristics of a capacitor

The basic unit of matter is the atom.

An atom has a nucleus consisting of protons which are positively charged and neutrons which have no electrical charge. Around the nucleus are several electrons which form part of the atom but are in orbit in the space surrounding the nucleus.

The electrons each have a minute negative electric charge and an atom which has its proper complement of electrons will have no overall electric charge because the negative polarity of the electron field exactly matches the positive polarity of the nucleus.

If however an electron manages to leave its host atom then that atom will have an overall positive charge. Conversely, an atom may acquire an electron from another source and thereby take on a negative charge.

The amount of charge contained in one electron is so small that isolated incidents of this kind have very little effect on surrounding conditions.

If the process is repeated many millions of times the effect does become significant and the charge developed can reach very large proportions.

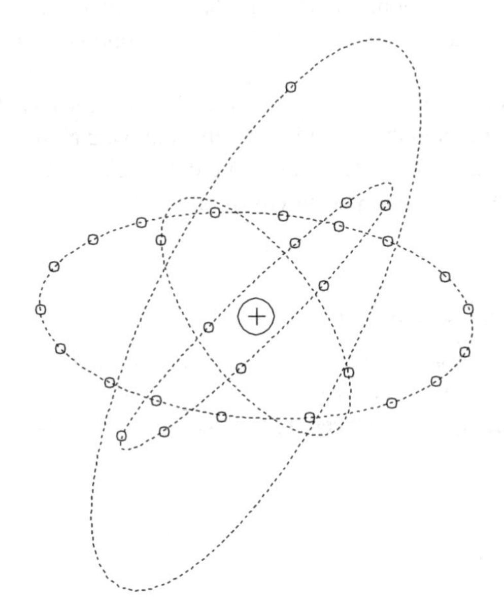

Figure 2.1 *A copper atom*

Electrostatic discharge

You may have noticed a painful discharge of electricity from your body on touching a metal object after having walked across a nylon carpet.

Touching the door handle of a car even after a short journey can produce a visible discharge of several millimetres.

This is all due to the migration of electrons caused by the movement of insulating surfaces, one upon the other. The voltages attained due to such circumstances can be very high and in situations where this can cause problems, steps must be taken to discharge the voltage before any harm can be done.

Persons working with delicate electronic components may have to wear wrist bands which are bonded to remove the charge to earth.

The process of loading and unloading flammable or explosive substances to and from road vehicles may involve bonding the vehicle to the installation before transferring the load.

Lightning is another form of static electricity in which the cumulative effect of convection over a long period of time causes an enormous build up of electric charge on the underside of a cloud mass which eventually leads to the massive discharge to earth we know as a lightning strike.

Static electricity was known to man long before cells, batteries and electromagnetic generating devices had been devised.

The word "electron" is derived from the Greek name for amber which, when rubbed with silk could be made to attract light objects.

Subsequent experiments over the centuries revealed that considerable amounts of electric charge could be accumulated from substances which reacted in the appropriate manner.

The fact that electricity can be stored in static form is of great importance and the effect is usefully employed in all kinds of equipment from minute electronic components to industrial applications of many hundreds of kilowatts.

The capacitor

The capacitor is a device which can store an electric charge. A capacitor usually consists of two conductive plates, separated from each other by a layer of insulation known as a dielectric.

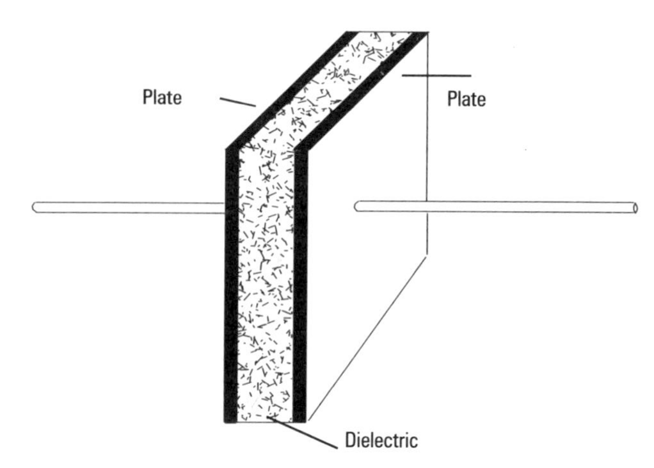

Figure 2.2 *The basic capacitor*

When connected to a source of electromotive force, a capacitor will accept a flow of electric current which will eventually cease when the device is fully charged.

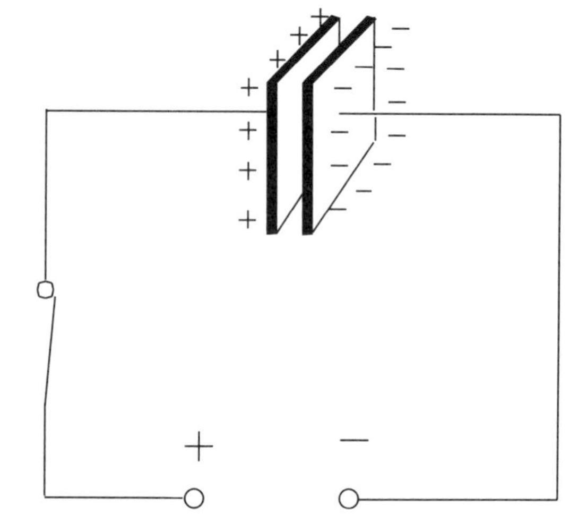

Figure 2.3

Charge

Symbol – Q Units – Coulombs Abbr. – C

$$Q = I\,t$$

A current of I amperes flowing for t seconds will result in a charge of Q coulombs.

So, for example, when a current of 200 mA flows into a capacitor for 40 mS a charge of 8 mC will be developed.

The amount of charge (the number of Coulombs) which can be stored in the capacitor will be determined by the area of the capacitor plates and the properties of the dielectric.

Capacitance

A capacitor which can store a charge of one Coulomb at a potential difference of one volt is said to have a CAPACITANCE of one FARAD (Abbr. – F)

$$C = \frac{Q}{U} \text{ (Farads) or}$$

$$Q = C\,U \text{ (Coulombs)}$$

The Farad is an extremely large unit, consequently the smaller unit microfarad (µF) is more commonly used.

Example

A capacitor of 200 µF which is fully charged at 200 volts holds a charge of

$$Q = C\,U$$

$$Q = 200 \times 10^{-6} \times 200$$

$$= 40 \text{ mC (milliCoulombs)}$$

Try this

A 240 µF capacitor is fully charged from a 100 V supply. Calculate the charge stored.

Capacitors in parallel and series

Parallel

When two or more capacitors are connected in **PARALLEL**, the effect is to increase the plate area under charge and therefore increase the capacitance in direct proportion.

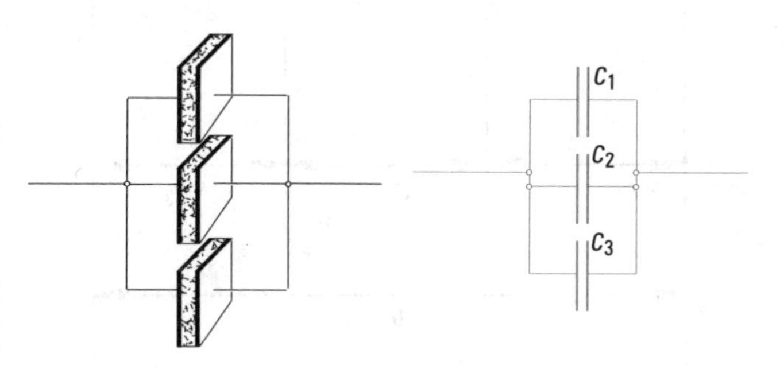

Figure 2.4

$$C_{(total)} = C_1 + C_2 + C_3$$

Example

A capacitor of 100 µF is connected in parallel with another of 150 µF and then a third of 50 µF. What is the final capacitance?

$$C = C_1 + C_2 + C_3$$

$$= 100 + 150 + 50$$

$$= 300 \text{ µF}$$

Try this

Four capacitors are connected in parallel to make up a total capacitance of 184 µF. Two capacitors are 32 µF each. A third is 48 µF. Calculate the value of the fourth.

Series

When capacitors are connected in **SERIES**, the effect on overall capacitance is quite different.

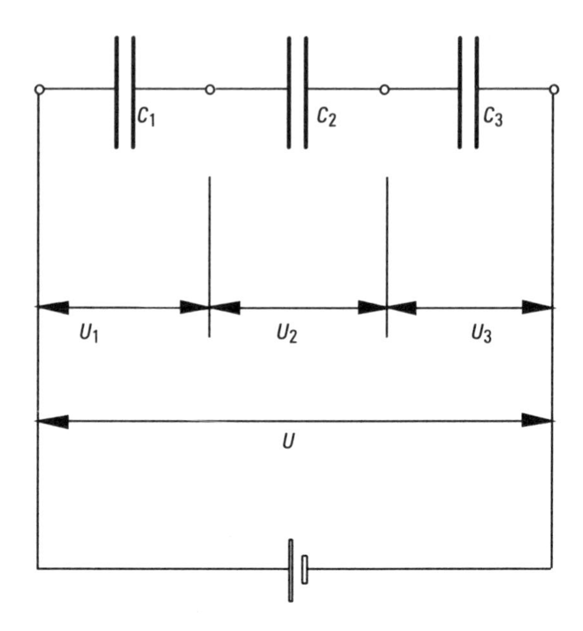

Figure 2.5

Because the three capacitors are connected in series, the charging current is the same in all three and flows for the same length of time.

i.e. $Q_1 = Q_2 = Q_3$

Indeed it can be said that the charge in the whole network is equal to that of any individual capacitor because the charging current in the whole series circuit is the same as any individual capacitor.

then $Q_{(total)} = Q_1 = Q_2 = Q_3$

It is obvious that the total voltage is the sum of the individual capacitor voltages.

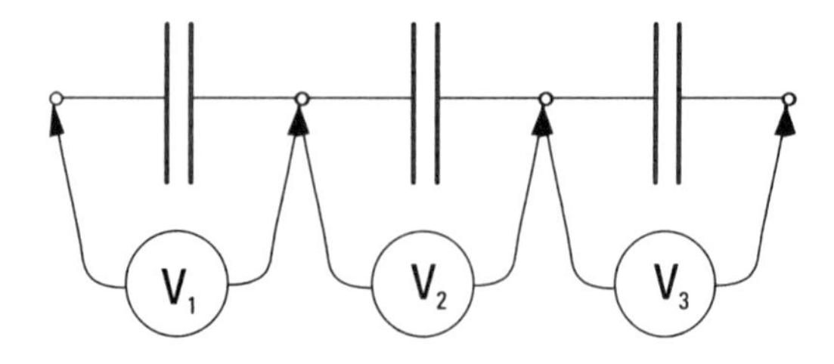

Figure 2.6

$$U_{(total)} = U_1 + U_2 + U_3$$

but $U = \dfrac{Q}{C}$

therefore

$$\frac{Q}{C_{(total)}} = \frac{Q}{C_1} + \frac{Q}{C_2} + \frac{Q}{C_3}$$

As Q is the same in each case, then we can divide each term by Q (in layman's terms the Qs cancel out)

and this gives us:

$$\frac{1}{C_{(total)}} = \frac{1}{C_1} + \frac{1}{C_2} + \frac{1}{C_3}$$

to take this a stage further:

$$C_{(total)} = \frac{1}{\frac{1}{C_1} + \frac{1}{C_2} + \frac{1}{C_3}}$$

Example

A series circuit consists of three capacitors of different capacitances as follows:

$$C_1 = 12\ \mu F$$
$$C_2 = 6\ \mu F$$
$$C_3 = 4\ \mu F$$

What is their combined capacitance?

$$C_{(total)} = \frac{1}{\frac{1}{12} + \frac{1}{6} + \frac{1}{4}}$$

$$= 2\ \mu F$$

It may be easier to think of combined capacitor circuits as being similar to, yet differing from, resistor networks as follows.

1. The combined value of several resistors in SERIES can be found by:

   $R_T = R_1 + R_2 + R_3$ etc.

   The combined value of several capacitances in PARALLEL can be found by:

   $C_T = C_1 + C_2 + C_3$ etc.

2. The combined value of several resistors in PARALLEL can be found by

$$R_t = \frac{1}{\frac{1}{R_1} + \frac{1}{R_2} + \frac{1}{R_3}} \text{ etc.}$$

The combined value of several capacitances in SERIES can be found by:

$$C_t = \frac{1}{\frac{1}{C_1} + \frac{1}{C_2} + \frac{1}{C_3}}$$

Try this

1. Four capacitors are connected in series. Their values are:

$$C_1 = 100\ \mu F$$
$$C_2 = 200\ \mu F$$
$$C_3 = 300\ \mu F$$
$$C_4 = 400\ \mu F$$

What is their combined capacitance?

2. What is the value of this arrangement?

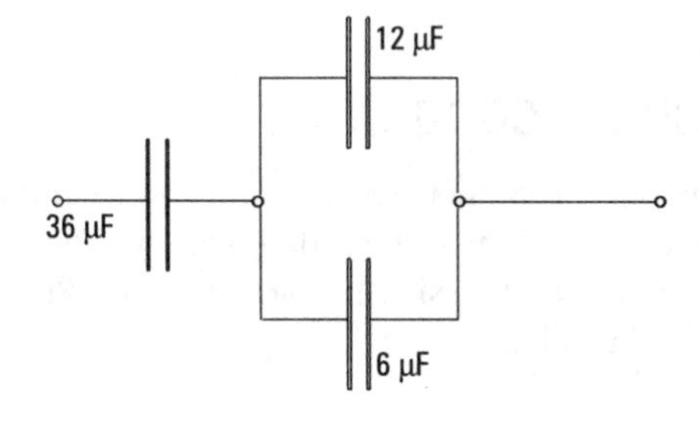

The energy stored in a capacitor

If a capacitor is charged at a constant rate of current the potential difference between the plates will increase at a constant rate.

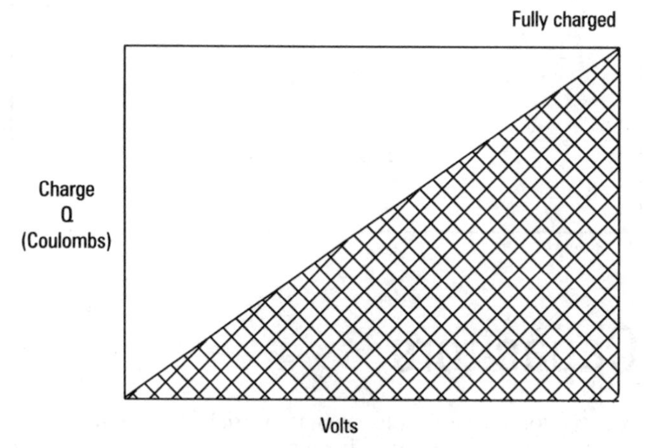

Figure 2.7 Voltage charge characteristics

The Charge is in Coulombs, which is Ampere Seconds

The potential difference is in volts

Multiplying these two quantities together gives

$U \times I \times t$ (Watt seconds)
or Joules (of energy)

But $Q = C\,U$

and if this is multiplied by the volts will give

$$C\,U^2$$

which is also an expression of energy in Joules.

As you can see, this graph, because it has a constant charging current, is a straight line which exactly bisects the figure and which has an area $C\,U \times U$.

Therefore; the energy stored in a capacitor can be found by

$$W = \tfrac{1}{2} C\,U^2 \text{ Joules}$$

Example

What is the energy stored in a 200 µF capacitor if the p.d. between the plates is 100 V?

$$W = \tfrac{1}{2} C\,U^2 \text{ Joules}$$

$$= \tfrac{1}{2} \times 200 \times 10^{-6} \times 100^2$$

$$= 1 \text{ Joule}$$

Try this

How much energy is stored in a 400 μF capacitor at a potential difference of 400 V?

Capacitors

A capacitor normally consists of two strips of aluminium foil, separated from each other by strips of paper, rolled up and inserted in a plastic or metal cylinder.

Plate area however is not the only factor which determines the capacitance of the device. There are three factors which affect capacitance:

1. The surface area of the plates
2. The distance between the plates
3. The insulating material between the plates, which is also known as the dielectric.

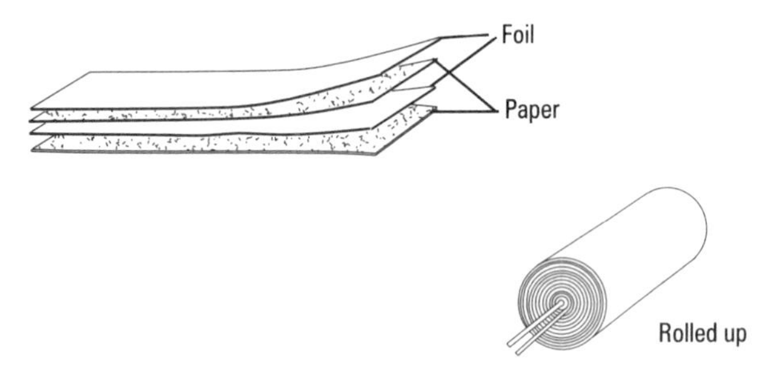

Figure 2.8 Foil type capacitor

1. The foil strips form the plates of the capacitor and consequently the larger the plate area, the greater the capacitance. Using very thin foil allows large plate areas to be accommodated in relatively small enclosures.
   In some types of capacitor, multiple plates are connected in parallel groups to increase the overall area.
2. For maximum capacitance, the plates must be as close as possible. For this reason, the insulating layer is kept very thin. A few thousandths of a millimetre of variation in the thickness of the dielectric can have a significant effect on the capacitance.
3. The dielectric has to resist the flow of electrons, sometimes at very high voltages and it is a matter of choice which material is to be used. Some types of capacitor have an air gap between the plates because air is quite effective as a dielectric but as can be seen from Table 2.1 there are other materials which are better.

Table 2.1

Material	Dielectric constant
air	1
aluminium oxide	10
glass	7.6
mica	7.5
mylar	3
paper	2.5
porcelain	6.3
quartz	5
tantalum oxide	11
vacuum	1

Table 2.2

Typical characteristics of some capacitors

Type of capacitor	Temperature range (°C)	Capacitance range	Voltage	
			a.c.	d.c.
paper foil	–30 to 100	0.01 F to 100 μF	250 to 630	
polyester foil	–40 to 100	100 pF to 2.2 μF	90 to 200	160 to 400
polystyrene	–40 to 70	100 pF to 0.6 μF		63 to 1000
ceramic – disc	–55 to 125	5 pF to 1 μF	63 to 250	63 to 10000
ceramic – monolithic	–55 to 125	0.001 μF to 10 μF		63 to 450
electrolytic – foil	–20 to 80	1 μF to 22000 μF		6.3 to 500
electrolytic – tantalum	–40 to 150	2.2 μF to 3500 μF		200 to 1000

Electrolytic capacitors

Electrolytic capacitors are made differently to other types and are very widely used throughout the industry. They provide more capacitance for their size than any other types but are not suitable for all applications.

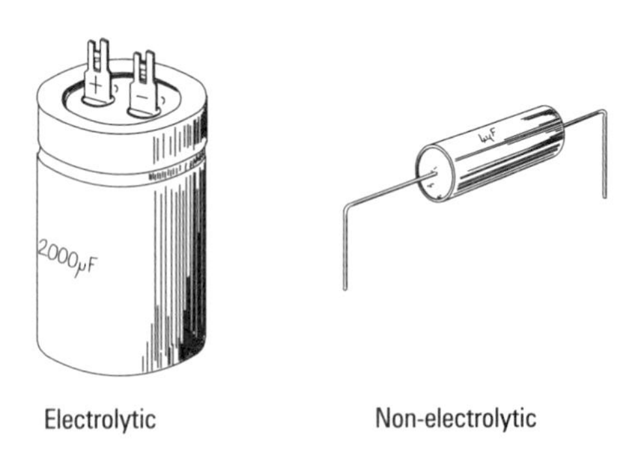

Figure 2.9

During the course of manufacture, the plates are subjected to an electrolytic process which oxidises one of the plates. As can be seen from Table 2.1, aluminium oxide is a very effective dielectric, being ten times better than air. The oxide dielectric is also very thin and it is this which helps to reduce the physical size of electrolytics when compared with other types.

Where an electrolytic capacitor is clearly marked + and - it must always be connected to a d.c. supply and the polarity must not be reversed as this will reverse the electrolytic process which deposited the oxide layer on the plate and the capacitor will be destroyed.

Tantalum electrolytic

Tantalum electrolytic capacitors are smaller than comparative aluminium oxide types and are made by a similar process which oxidises tantalum foil electrodes. They are widely used for electronic applications and with their characteristic "bead" shape can be easily identified on a circuit board.

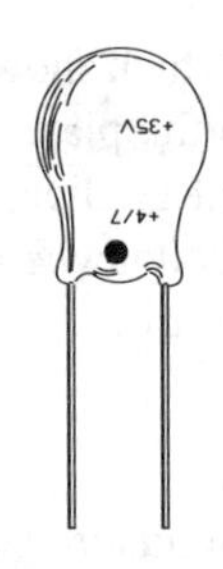

Figure 2.10 Tantalum

Other types

Plastic film dielectrics are formed from a variety of materials such as polypropylene, polystyrene and polycarbonate and are usually manufactured by a process which deposits the metallic layer on the film. The finished capacitor is is usually encapsulated in an epoxy resin enclosure. These are very commonly used in electronic circuits and can be identified by their small box-like appearance.

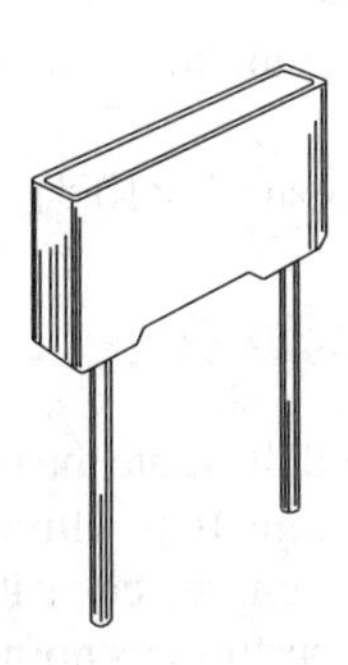

Figure 2.11 Plastic film

Silver-mica capacitors are used in high frequency applications where high stability and accuracy are essential. They are made by depositing silver on layers of mica. They are expensive by comparison with other types.

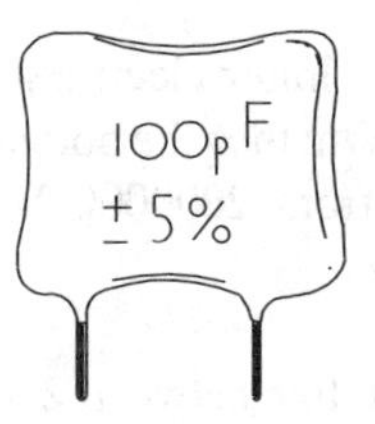

Figure 2.12 Silver-mica

Ceramic capacitors consist of multiple layers of metallised ceramic discs. They are usually low-value, small size capacitors and are used for high frequency circuits.

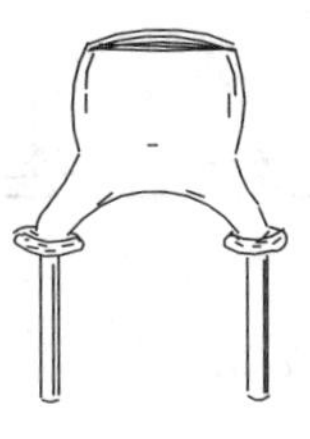

Figure 2.13 Ceramic

Dielectric strength

The voltage which can be applied between the plates of a capacitor is not infinite. There is a point at which any insulator will break down and the material will eventually permit a current to flow between the plates.

Since dielectric materials are chosen for their ability to withstand high voltages without breaking down, then this property is of considerable importance.

The potential difference required to break down an insulator is termed its **DIELECTRIC STRENGTH** and is expressed in Volts per metre.

The distance between the plates is never very great and when measured in metres is only a tiny fraction of a metre and even then, the voltage between the plates can be quite substantial. This means that the dielectric strength is expressed in very large numerical terms. For practical purposes, the dielectric strength may be given in kilovolts per millimetre and even then, the quantity can be quite large.

Potential gradient

If the ability to withstand an electrical breakdown is to be expressed as a constant with respect to any particular dielectric then a comparison between materials can be made in these terms.

For example if the maximum electric stress for an air dielectric is to be taken as 2 MV/m then the potential difference between the plates will fall from 2000000 V to zero Volts over a distance of one metre.

By drawing a straight line between 2×10^6 and zero then the appropriate voltage can be found for any intermediate point.

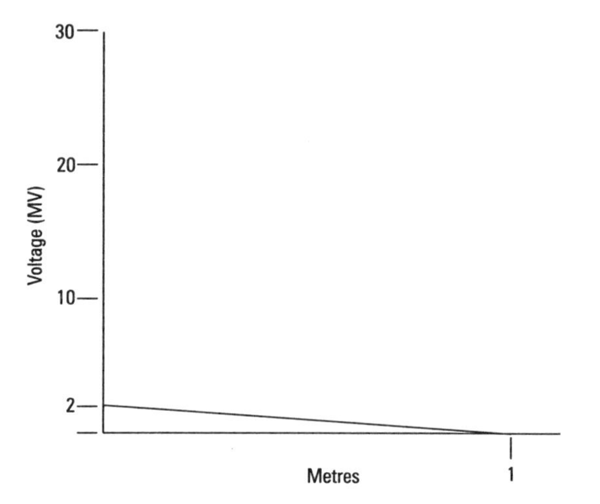

Figure 2.14 *Air dielectric*

To take the example of a paper dielectric, the maximum stress may be found to be of the order of 5 MV/m and a straight line drawn from 5000000 V to zero volts will serve the same purpose but in this case it will be slightly steeper.

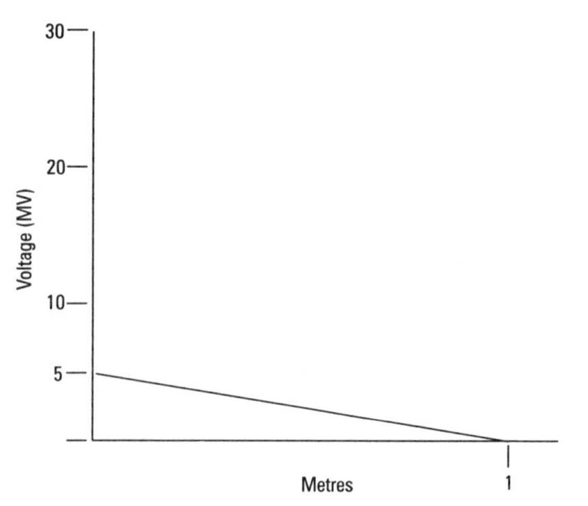

Figure 2.15 *Paper dielectric*

Mica, being a dielectric material which is in common use has maximum electric stress of 100 MV/m and the line would be fifty times as steep as that drawn for air.

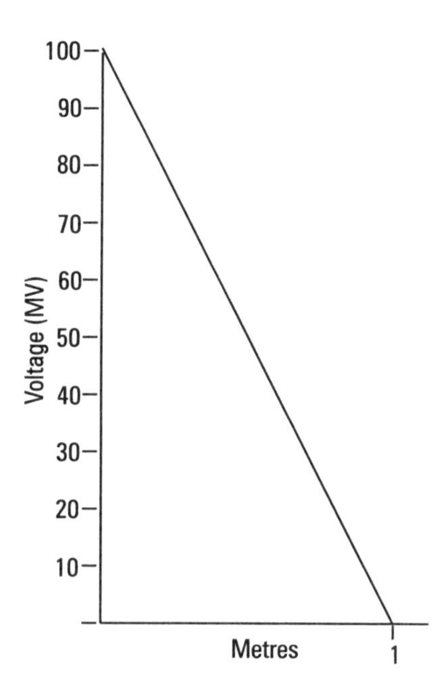

Figure 2.16 *Mica dielectric*

The steepness of the line which shows the maximum stress for a given distance between the plates for a specific material is referred to as its potential gradient and this is an important feature to be taken into account when comparing the properties of different types of capacitor.

Example

If the potential gradient of an air-spaced capacitor is given as 2 MV/m, what is the maximum voltage to be applied between the plates if the air gap is 0.7 mm?

Maximum possible voltage

$= \text{Potential gradient (V/m)} \times \text{distance in metres}$

$= (2 \times 10^6) \times (0.7 \times 10^{-3})$

$= 1400 \text{ V } (1.4 \text{ kV})$

Now compare this with a polythene capacitor having a dielectric strength of 40 MV/m but the same dielectric thickness.

Maximum possible p.d.

$= \text{Potential gradient} \times \text{thickness}$

$= (40 \times 10^6) \times (0.7 \times 10^{-3})$

$= 28000 \text{ V } (28 \text{ kV})$

These two examples will illustrate the importance of dielectric strength in capacitor design. If the ultimate breakdown voltage is exceeded for any reason, the capacitor may be completely destroyed and cease to perform its normal function. Particular care must be taken with electrolytic capacitors as the dielectric strength is dependent on correct connection and the device may be completely unsuitable for a.c. applications.

Try this

Calculate the maximum possible voltage for a capacitor with a potential gradient of 60 MV/m and a dielectric thickness of 0.08 mm.

The time constant

If the current which is charging a capacitor is controlled by a resistor, it will rise initially at a constant rate and the result of this is that the capacitor voltage will also start to rise uniformly. If this rate of increase continued until the capacitor was fully charged the time taken would be CR seconds.

This is known as the **TIME CONSTANT** and is given the symbol τ (TAU).

Time Constant τ = CR seconds

Example

What is the time constant for a 100 µF capacitor which is being charged through a 5000 Ω resistor?

$$\text{Tau } \tau = CR$$
$$= 100 \times 10^{-6} \times 5 \times 10^{3}$$
$$= 0.5 \text{ seconds}$$

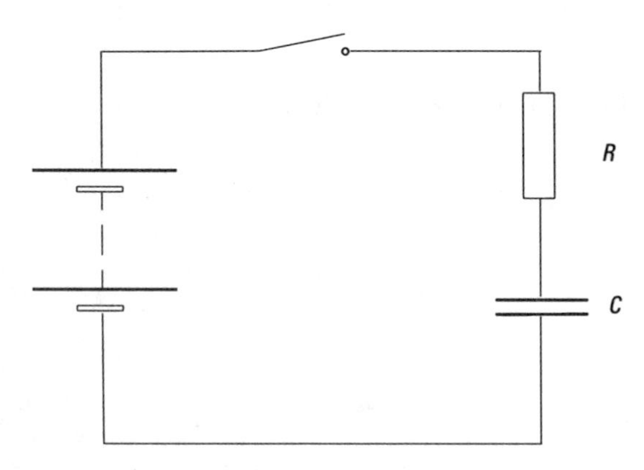

Figure 2.17 *"CR" circuit*

What happens in practice is that as the capacitor p.d. increases the current reduces and therefore the rate of voltage rise "flattens out" as the capacitor p.d. approaches that of the supply.

Calculate the time constant (τ = CR)

This growth curve can then be produced graphically by the following method. Draw a straight line from the origin to a point on the maximum voltage line at *t* seconds from the start of the graph (Figure 2.18).

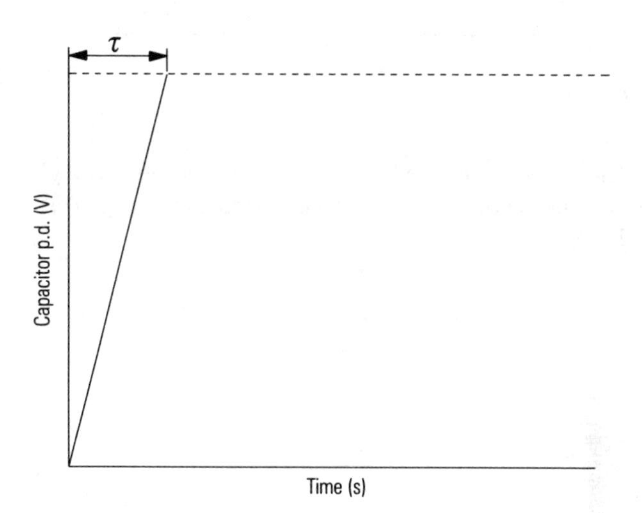

Figure 2.18

Choose a point on this line (say one third of the way up), measure horizontally, τ seconds and then vertically upwards until you reach the top line. Join this to form a triangle as shown in Figure 2.19. Similarities can be seen in this and the current growth in an inductor in Chapter 1.

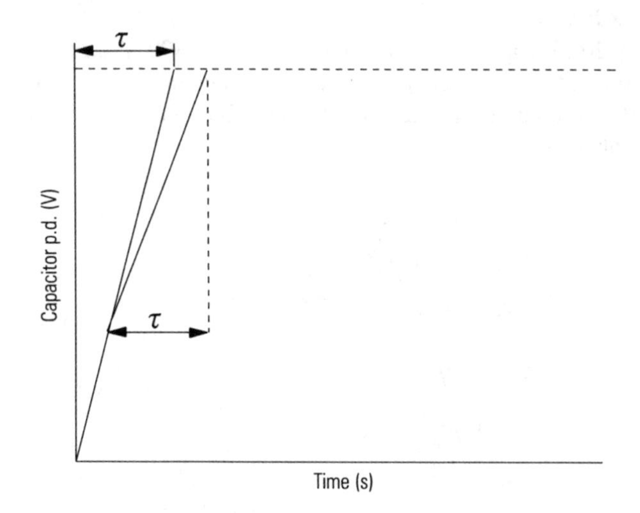

Figure 2.19

Keep repeating the process until you have a smooth curve which merges with the top line (Figure 2.20).

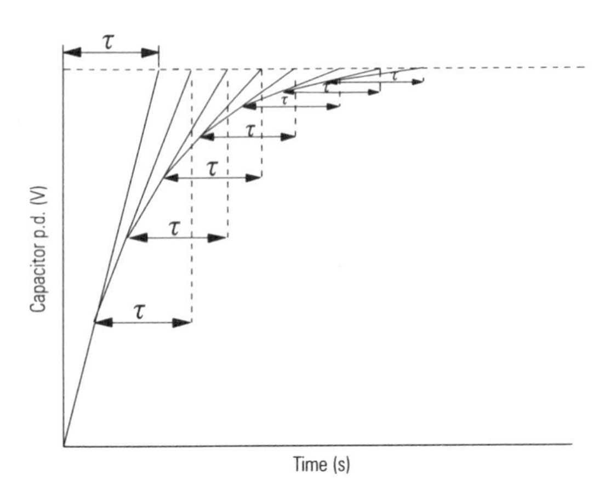

Figure 2.20

This graph can be used to determine the voltage after a given time, or the time taken to reach a particular voltage (Figure 2.21).

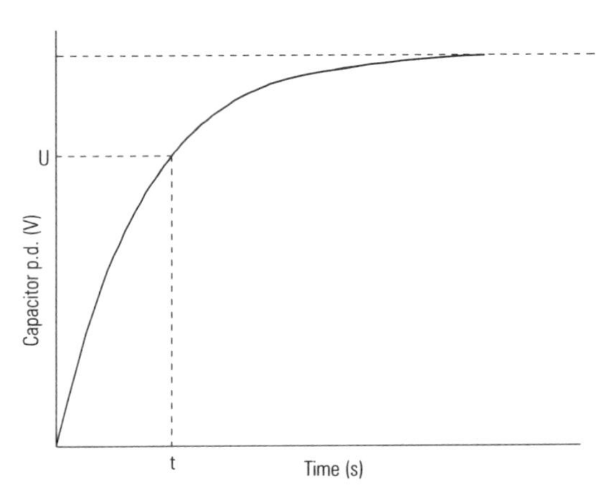

Figure 2.21

Example

A 20 μF capacitor is charged to a p.d. of 800 V through a 100 kΩ resistor. Calculate the time constant of the circuit and derive a curve in order to determine the time taken for the voltage to reach 500 V.

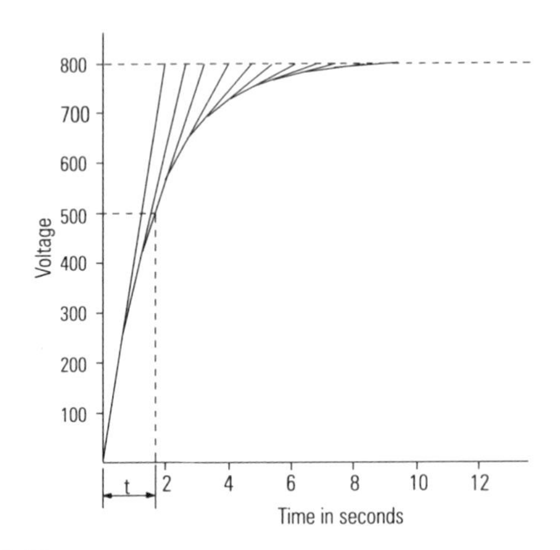

Figure 2.22

From the graph, *t* is approximately 1.8 seconds

Current during the charge of a capacitor

As the voltage increases when a capacitor charges, the current in the circuit decreases. When the capacitor is first switched on the current reaches a maximum value. As the flow of electrons to the plates becomes less the current decreases. The curve of the current against time can be plotted in a similar way to the voltage but it will start at a maximum of

$$I = \frac{U}{R} \quad \text{and decrease.}$$

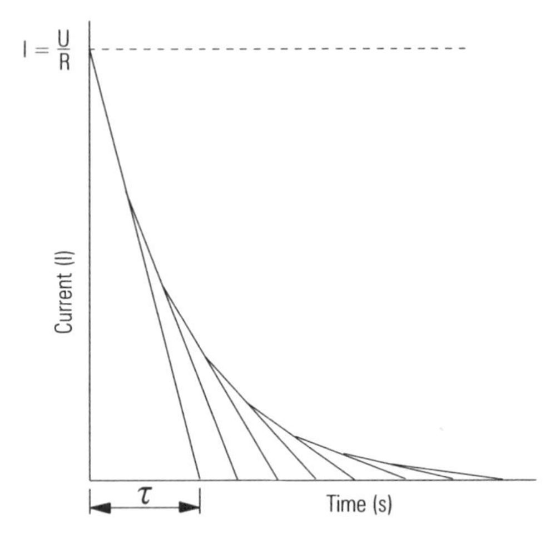

Figure 2.23

Try this

Using Figure 2.22 determine

(a) the voltage after 3 seconds charge

(b) the time it takes to charge to 750 V.

Discharge of a capacitor

When a charged capacitor is discharged into a resistor as in Figure 2.24, the time constant can be found by the same method but in this case the voltage across the capacitor will be seen to fall as shown in Figure 2.25.

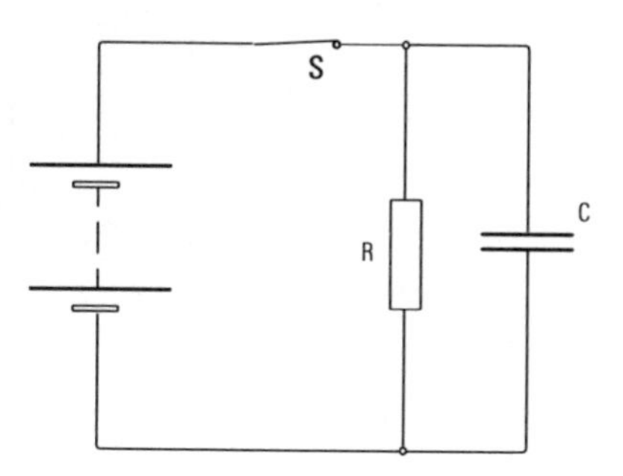

Figure 2.24

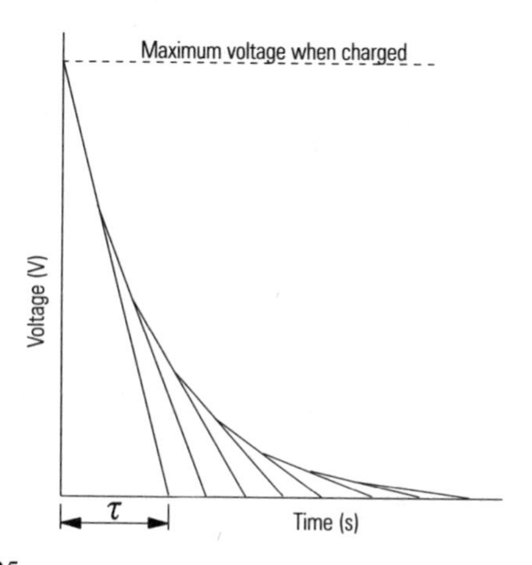

Figure 2.25

Example

A 200 μF capacitor is connected to a 100 V d.c. supply. When the supply is switched off the capacitor discharges into a 1 MΩ resistor. What is the voltage remaining in the capacitor 100 seconds after the supply is switched off.

$\tau = CR$

$= 200$ seconds

Figure 2.26

The voltage is approximately 57 volts.

Try this

Using Figure 2.26 determine

(a) the voltage after 420 seconds

(b) the time it takes to discharge to 30 V.

Discharge resistors

It is not uncommon to find a capacitor with a resistor connected across its terminals. The reason for this is that once a capacitor is charged, it can hold its charge for a considerable length of time. Any person coming into contact with the capacitor could receive a painful electric shock long after the mains supply has been isolated.

Accidents have resulted from persons receiving an electric shock from capacitors when working on ladders or scaffolds causing them to step back or lose their grip and suffer injury from falling.

A discharge resistor fitted to the capacitor will safely dissipate the charge and help to prevent accidents and injuries.

The choice of discharge resistor will depend on the circumstances, but it must have a low enough value to discharge the capacitor quickly enough to prevent inadvertent electric shock. At the same time it must not be so low as to interfere with the correct operation of the circuit.

It must be remembered that any two conductive surfaces which are separated from each other by an insulating medium can form a capacitor.

It is frequently found that the capacitance between the cores of a multicore cable can produce unexpected effects, particularly when low values of current and voltage are being used.

Where long cables are used for control circuitry in an a.c. system care should be taken to ensure that alternating current does not continue to flow due to the capacitance of the cable when the circuit has been opened.

Another situation from which danger can arise is after the isolation of long overhead conductors. The overhead cable can and will hold a charge, sometimes at a very high voltage between the conductors and earth.

When isolating such a system it is proper procedure to connect the conductors to earth and keep them earthed while work is being carried out. Most H.V. circuit breakers have an earthing facility, i.e. ON – OFF – EARTH, which will ensure safety during maintenance operations.

It would not be considered good practice for anyone, however brave, to approach a high voltage overhead cable with an earthed "croc clip".

A.C. applications of capacitors

Having seen that the capacitor can hold a charge derived from a d.c. source and then deliver that charge back into the circuit in a slow (or sometimes not so slow) measured manner, one is inclined to think of such devices only in these terms.

There are however many a.c. circuits which rely on capacitors for their proper operation.

When connected to an a.c. source the current flows into the device during the first half cycle. At the start of the second half cycle, the polarity changes and the charge which has been accumulated discharges back into the circuit and then the device charges up again but this time with the opposite polarity. So the sequence goes on indefinitely. No current has passed through the dielectric but an alternating current has shunted up and down the conductor without hindrance.

One important feature of this behaviour is that the voltage across the capacitor is dependent on the current, i.e. current first, voltage afterwards.

This means that the current will lead the voltage, up and down, thus giving the effect of a leading power factor.

Power factor correction capacitors handling many hundreds of amperes are to be found in industrial installations where a lagging power factor, caused by heavy motor loads would otherwise cost the consumer dearly in tariff surcharges. This is covered in more detail in Chapter 4.

The single phase capacitor-start motor needs the phase shifting effect of the capacitor to make it rotate, and discharge lighting gear normally incorporates a capacitor to counteract the effect of the inductor on the load current.

Exercises

1. A series circuit consists of three capacitors of 10 μF, 20 μF and 40 μF respectively connected to a 400 V supply. Calculate:
    (a) single capacitance equivalent to replace the three
    (b) total energy stored across all three capacitors
    (c) voltage drop across each capacitor
    (d) time constant to discharge the circuit when a 1 MΩ resistor is connected across the charged capacitor.

2. A series circuit containing two capacitors 20 μF and 80 μF respectively and a 50 kΩ resistor are connected across a 300 V d.c. supply.
   (a) Calculate the time constant of the circuit.
   (b) Construct a graph showing the current flowing in the circuit with respect to time.
   (c) Use the graph to estimate the time for the current to fall to 2 mA.
   (d) Calculate the voltage across each capacitor when the energy stored in the circuit is 20 mJ.

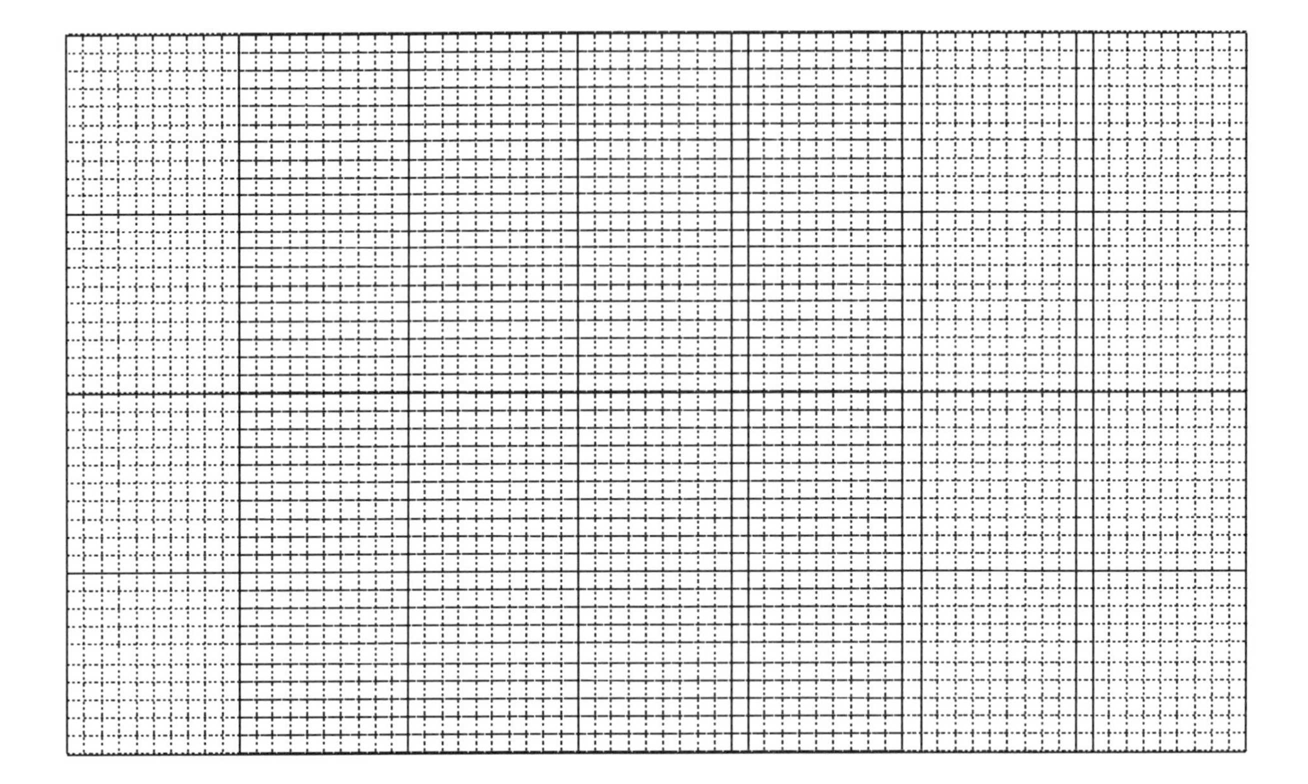

3

Circuit Theory

Complete the following to remind yourself of some important facts on this subject that you should remember from the previous chapter.

1. When considering parallel and series networks, what is the significant difference between resistors and capacitors?

2. The layer of insulating material between the plates of a capacitor is called the

3. A high value resistor, permanently connected across the terminals of a capacitor, is there for what purpose?

On completion of this chapter you should be able to:

- determine the effects on voltage and current in series and parallel circuits with only resistance present
- calculate the voltages and currents in a circuit using Kirchhoff's Laws
- determine the value of unknown resistors using bridge circuits
- determine the impedance of circuits containing reactance and resistance
- draw scaled phasor diagrams to determine unknown quantities
- determine the currents and their relationships in three-phase loads

Resistor networks

There are very few problems which fall within the broad area of circuit theory which cannot be solved by the application of Ohm's Law. When considering an electric current, a potential difference or an impedance of any kind, if any two of these variables can be established then the third can be found.

Electrical circuits of alarming complexity can be and are devised but in the vast majority of cases, these can be broken down into a network of sub-circuits which can be dealt with individually until the workings of the whole system and all its component parts are fully understood.

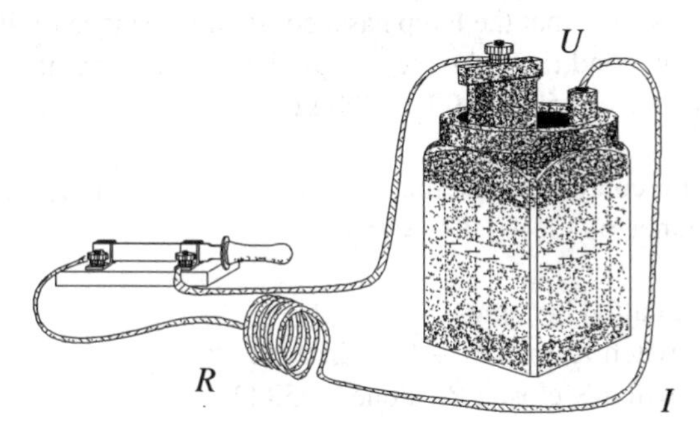

Figure 3.1

It is often necessary to look at a circuit as a complete item before searching through its insides to find the answer to a particular problem.

Example

The lamp filament, in Figure 3.2, has to be supplied with a current of 20 mA when switch X is open and 40 mA when switch X is closed. The supply is 50 V d.c. and the lamp has a resistance of 300 Ω.

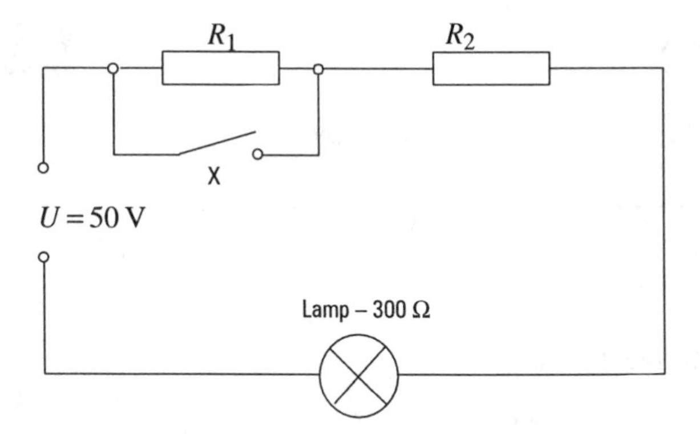

Figure 3.2

What are the resistor values?

The problem could be solved by this method;-

1) With switch X open

$$R = \frac{U}{I}$$

$I = 0.02$ A
$U = 50$ V

$$R = \frac{50}{0.02}$$

$= 2500\ \Omega$

2) With switch X closed

$$R = \frac{U}{I}$$

$I = 0.04$ A
$U = 50$ V

$$R = \frac{50}{0.04}$$

$= 1250\ \Omega$

If we assume that the lamp has a constant resistance of 300 Ω then in condition (1) the remainder of the circuit has a resistance of 2500 – 300 = 2200 Ω

and in condition (2) the remainder of the circuit has a resistance of 1250 – 300 = 950 Ω

The situation becomes

1) switch X open $R_1 + R_2 = 2200\ \Omega$
2) switch X closed R_2 alone = 950 Ω

The difference between the two is the resistance of R_1 which is 2200 – 950
i.e. 1250 Ω

Check it out.

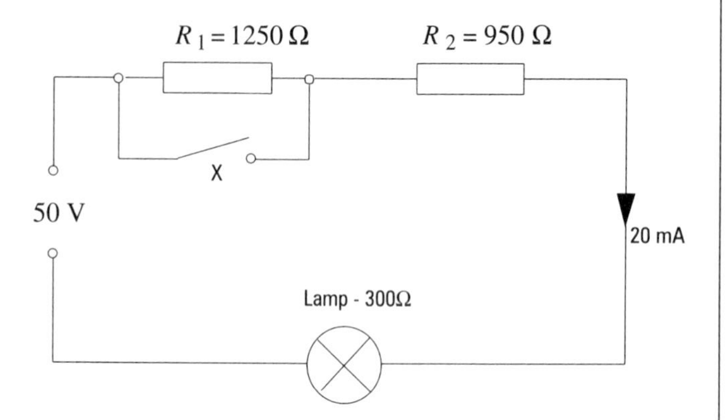

Figure 3.3

1. Switch X open

$R_{(total)} = 1250 + 950 + 300$

$= 2500\ \Omega$

$$I = \frac{U}{R} = \frac{50}{2500}$$

$= 0.02$ A (20 mA)

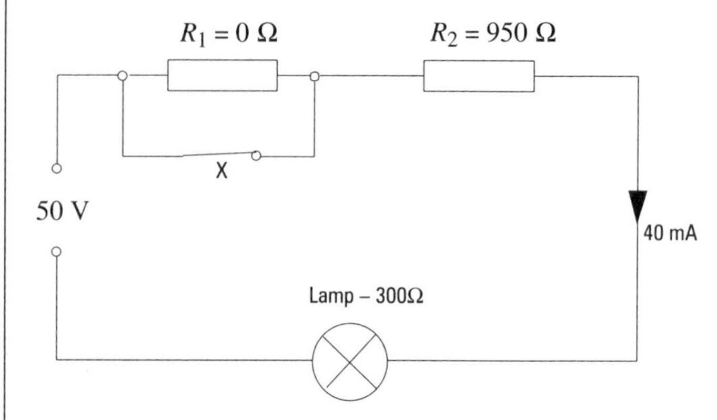

Figure 3.4

2. Switch X Closed

$R_{(total)} = 950 + 300$

$= 1250\ \Omega$

$$I = \frac{U}{R} = \frac{50}{1250}$$

$= 0.04$ A (40 mA)

So now you will begin to see what you are up against.

Example

Now we'll devise a small variation on the same theme.

Suppose we want to control the lamp's brightness with a parallel resistor rather than a series one, which will make the circuit look like this:

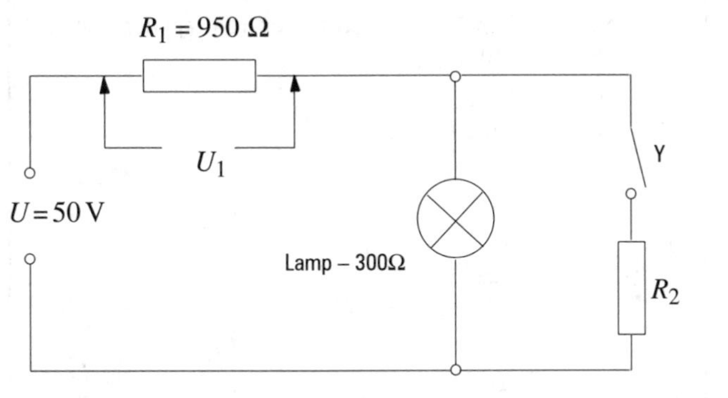

Figure 3.5

The conditions now become:

1) Switch Y open, total circuit resistance $= 1250\ \Omega$
2) Switch Y closed, lamp current $= 20$ mA

Assuming again that the lamp remains constant at 300 Ω this leaves us with the resistor network as follows

1) R_1 alone = 1250 – 300 = 950 Ω (simple enough)
2) R_1 in series with lamp and R_2 in parallel (that's a different story)

Let's go back and look at what is required:

1) lamp glowing brightly with 40 mA going through the filament
2) lamp dim with 20 mA going through the filament

OK – What does Ohm's Law tell us about these two conditions?

1) The potential difference across the lamp is

$$U = I \times R = 0.04 \times 300 = 12 \text{ V}$$

when glowing brightly

2) The potential difference across the lamp is

$$U = I \times R = 0.02 \times 300 = 6 \text{ V}$$

when dimmed

This means that the voltage drop across R_1 has got to be 50 – 12, i.e. 38 V in condition (1) and 50 – 6 i.e. 44 V in condition (2).

1) $R_1 = 950\ \Omega$, $I = 0.04$ A

$$U = I \times R = 0.04 \times 950 = 38 \text{ V (so that's OK)}$$

2) $R_1 = 950\ \Omega$, $U_1 = 44$ V

This requires a current of

$$I = \frac{U_1}{R_1} = \frac{44}{950} = 0.0463 \text{ A}$$

Remember Ohm's Law?

Well let's apply it some more.

The p.d. across the lamp is 6 V.

The current in the lamp is 0.02 A

The p.d. across the parallel resistor is the same as the lamp i.e. 6 V.

The current through the resistor is what's left after you take away the lamp current:

i.e. 0.0463 – 0.02 = 0.0263 A

See what is happening?

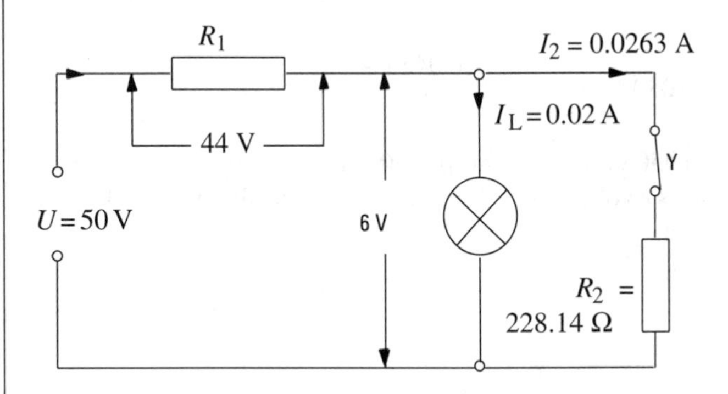

Figure 3.6

So the unknown resistor has a resistance of

$$R = \frac{U}{I}$$

$$= \frac{6}{0.0263}$$

$$= 228.14\ \Omega$$

And nothing more than Ohm's Law was used in the solution.

Check it out

The combined resistance of the lamp and parallel resistor can be calculated by the "product over sum method".

i.e. $$R_{(total)} = \frac{(R_a \times R_b)}{(R_a + R_b)}$$

where

R_a (lamp) $= 300\ \Omega$

R_b (resistor) $= 228.14\ \Omega$

$$= \frac{300 \times 228.14}{300 + 228.14}$$

$$= \frac{68442}{528.14}$$

$= 130\ \Omega$ (very nearly)

This gives a total circuit resistance of

$950 + 130 = 1080\ \Omega$

From a 50 V supply, this draws a current of

$$\frac{50}{1080} = 0.0463\ \text{A}$$

In a 950 Ω resistor, this produces a volts drop of 44 V, which leaves a voltage of 6 V across the lamp and this was the desired effect.

Try this

1. In the diagram below, with switch Y open, calculate the resistances of

   (a) R_1
   (b) the lamp
   (c) R_3

2. When switch Y is closed the current I_1 increases to 120 mA. Calculate
   (a) current I_R
   (b) resistance R_2

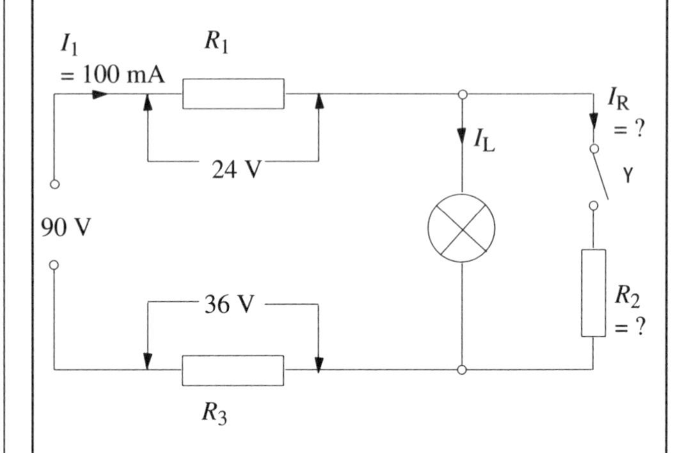

Kirchhoff's Laws

There are two laws attributed to the German physicist Gustav Kirchhoff these are very simple but can be useful when dealing with problems in electrical circuitry.

1st Law

The algebraic sum of the currents flowing in and out of a junction in a circuit is ZERO.

Example

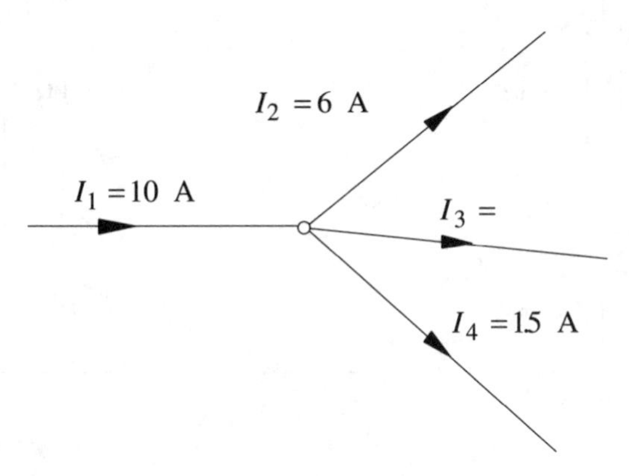

Figure 3.7 Currents flowing to and from a point.

Currents flowing into the junction are **positive**.

Currents flowing from the junction are **negative**.

$$I_1 - I_2 - I_3 - I_4 = 0$$

$$\therefore I_1 - I_2 - I_4 = I_3$$

$$\therefore I_3 = 10 - 6 - 1.5$$

$$= 2.5 \text{ A}$$

2nd Law

In any closed loop in a circuit, the algebraic sum of the applied e.m.f.'s and the voltage drops is ZERO.

Example

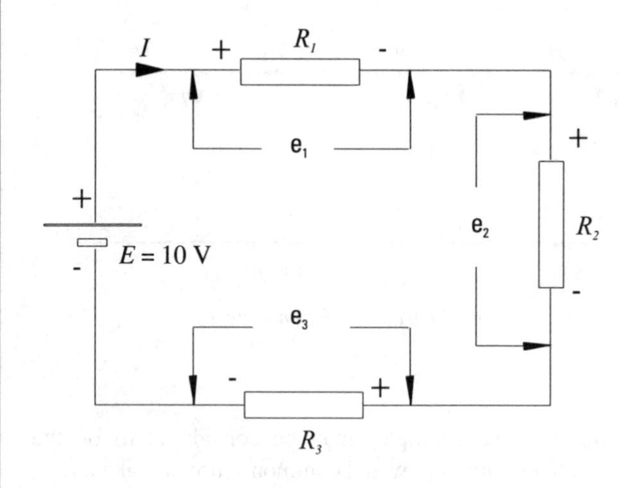

Figure 3.8 Sum of e.m.f. and p.d.s

Given that $R_1 = 2\ \Omega$

$R_2 = 1\ \Omega$

$R_3 = 1\ \Omega$

$R_{(total)} = 4\ \Omega$

$$I = \frac{E}{R} = \frac{10}{4} = 2.5 \text{ A}$$

("E" is used to denote e.m.f. and "e" a potential difference.)

and

$$e_1 = I R_1$$

$$e_2 = I R_2$$

$$e_3 = I R_3$$

Potential differences acting anticlockwise are **negative**.

$$E - I R_1 - I R_2 - I R_3 = 0$$

$$10 \text{ V} - 5 \text{ V} - 2.5 \text{ V} - 2.5 \text{ V} = 0$$

A good example of Kirchhoff's first law can be found in a d.c. 3-wire distribution system such as in the example given.

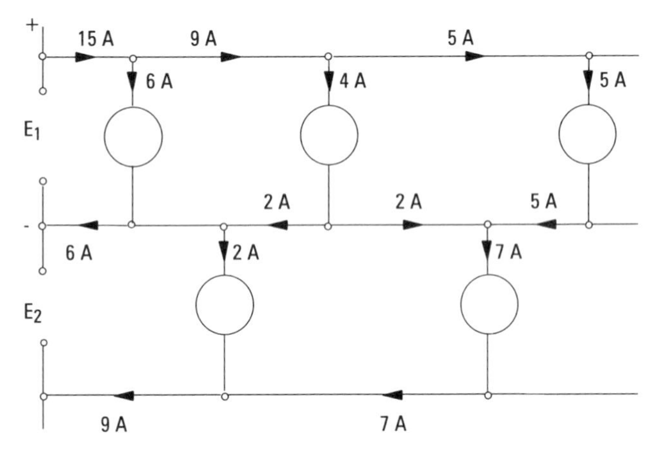

Figure 3.9 *DC 3-wire distribution system*

Of course, these examples may be considered to be the application of Ohm's Law and common sense but take a look at the next one and see how Kirchhoff's laws can be put to use in a more complex setting.

Example

As you can see, the network has two sources of e.m.f. and three closed loops and a straightforward solution by Ohm's Law is not possible.

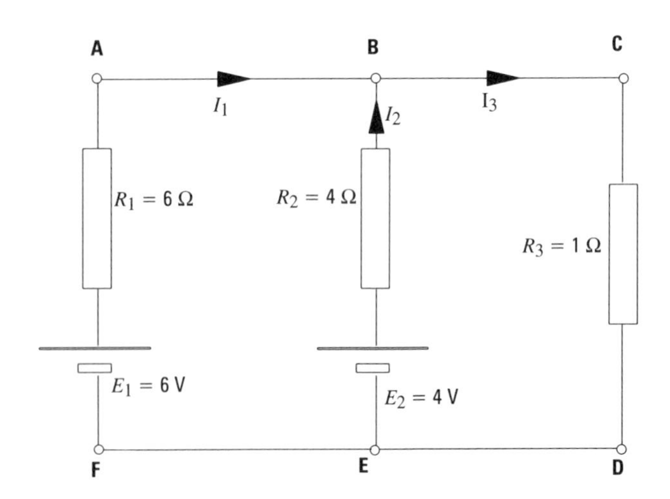

Figure 3.10

On inspection you will see that Kirchhoff's first law applies to I_3 and this can be expressed as $(I_1 + I_2)$.

$E_1 = 6$ V
$E_2 = 4$ V
$R_1 = 6\ \Omega$
$R_2 = 4\ \Omega$
$R_3 = 1\ \Omega$

Loop 1 **ABEF**
Loop 2 **BCDE**
Loop 3 **ACDF**

Taking loop **ACDF** (the one that goes right round the outside) the algebraic sum of the applied e.m.f. and the voltage drops is **ZERO**.

Or alternatively, the sum of the voltage drops is equal to the applied emf, which amounts to the same thing.

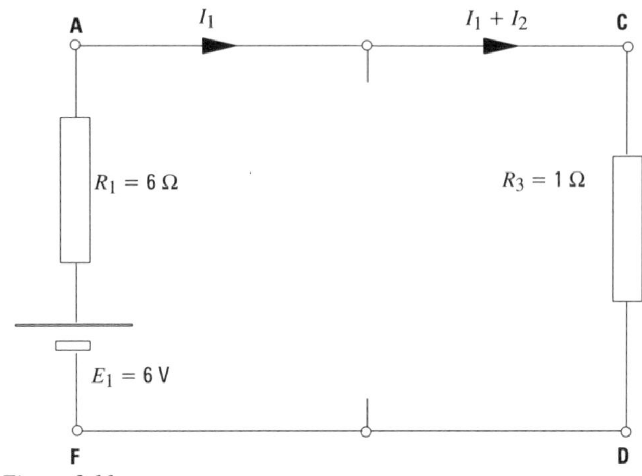

Figure 3.11

For this loop

$$6\text{ V} = I_1 \times R_1 + (I_1 + I_2) \times R_3$$

$$= 6 I_1 + 1 I_1 + 1 I_2$$

$$6 = 7 I_1 + I_2$$

Now take a look at loop **BCDE**

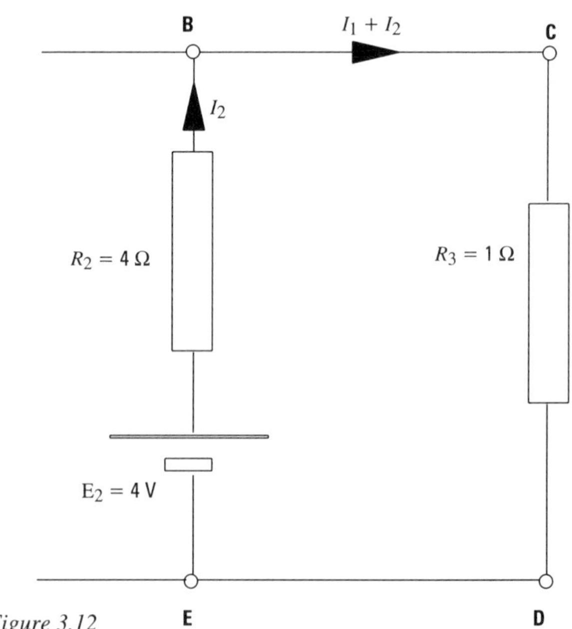

Figure 3.12

The sum of the voltage drops is equal to the applied e.m.f.s

hus

$$4\text{ V} = 4 \times I_2 + 1 \times (I_1 + I_2)$$

$$= 4 I_2 + I_2 + I_1$$

$$4 = I_1 + 5 I_2$$

Putting these two expressions together in the form of a simultaneous equation gives

$7 I_1 + I_2$	$= 6$	Equation 1
$I_1 + 5 I_2$	$= 4$	Equation 2

Multiply equation (1) by 5 giving

$$\begin{array}{lll} 35 I_1 & + 5 I_2 & = 30 \quad \text{and subtract equation 2} \\ I_1 & + 5 I_2 & = 4 \\ \hline 34 I_1 & & = 26 \end{array}$$

Thus $I_1 = \dfrac{26}{34}$

$$I_1 = 0.764\text{ A}$$

Substitute 0.764 for I_1 in equation 2.

$$0.764 + 5 I_2 = 4$$

hen $5 I_2 = 4 - 0.764 = 3.236$

herefore $I_2 = 0.647\text{ A}$

The whole solution becomes

$I_1 = 0.764\text{ A}$
$I_2 = 0.647\text{ A}$
$I_3 = 1.411\text{ A}$

Try another **Example**.

Consider Loop **ABEF**, and assuming that the current will flow in a clockwise direction, the sum of the voltage drops equals the applied e.m.f.

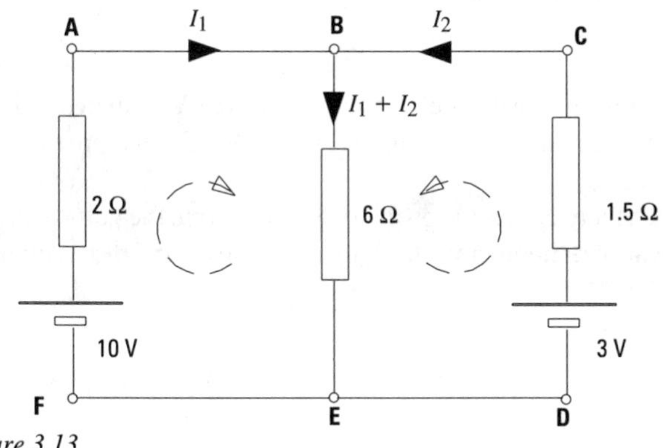

Figure 3.13

$$10\text{ V} = 2 I_2 + 6(I_1 + I_2)$$

$$= 2 I_1 + 6 I_1 + 6 I_2$$

$$8 I_1 + 6 I_2 = 10 \quad \text{(equation 1)}$$

Now take loop **CBED** assuming that the current is flowing in an anticlockwise direction

$$3\text{ V} = 6(I_1 + I_2) + 1.5 I_2$$

$$= 6 I_1 + 6 I_2 + 1.5 I_2$$

giving

$$6 I_1 + 75 I_2 = 3 \quad \text{(equation 2)}$$

To eliminate I_1, multiply the whole of equation (1) by 3 and equation (2) by 4 giving

$$\begin{array}{lll} 24 I_1 & + 18 I_2 & = 30 \quad \text{and subtract equation 2} \\ 24 I_1 & + 30 I_2 & = 12 \\ \hline & -12 I_2 & = 18 \end{array}$$

Thus $I_2 = -1.5$ A (more about the minus sign later)

Substitute for I_2 in equation (1).

$$8 I_1 + 6 I_2 = 10$$
$$8 I_1 + 6 \times (-1.5) = 10$$

this is the same as saying

$$8 I_1 - 9 = 10$$

or $8 I_1 = 19$

giving $I_1 = 2.375\text{ A}$

$I_2 = -1.5\text{ A}$

If a solution comes out with a minus sign, all this means is that the direction of the current flow is opposite to that first assumed. i.e. the current in the loop was clockwise, not anticlockwise as at first assumed.

The effect of this is that the current $(I_1 + I_2)$ becomes

$I_3 = 2.375 - 1.5$
$I_3 = 0.875\text{ A}$

Try this

The diagram shows a battery charger which supplies current to a load and charges a battery at the same time. By using Kirchhoff's Law and the resulting simultaneous equations, determine

(a) the output current of the charger
(b) the charging current of the battery
(c) the current in the load resistor
(d) the voltage at the load resistor terminals

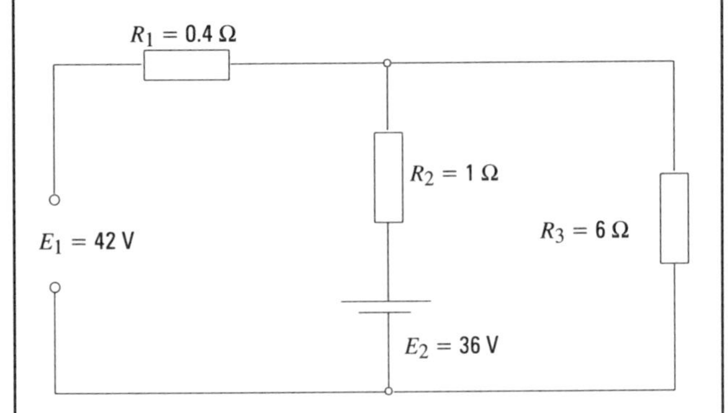

Ring distribution

A ring distribution conductor is a single loop conductor fed at one point but supplying loads from several places along its length. It is a simple matter to establish how much current is entering the ring and how much is leaving but it can be a tricky task determining the actual current in each individual section of the ring.

Example

Take the ring main cable shown in Figure 3.14. The resistance of the cable core is 0.6 m Ω/m. It is fed at **A** and has loads at **B** and **C** of 200 A and 300 A respectively. The current coming in at **A** is obviously 500 A but how much of this is going right and how much is going to the left.

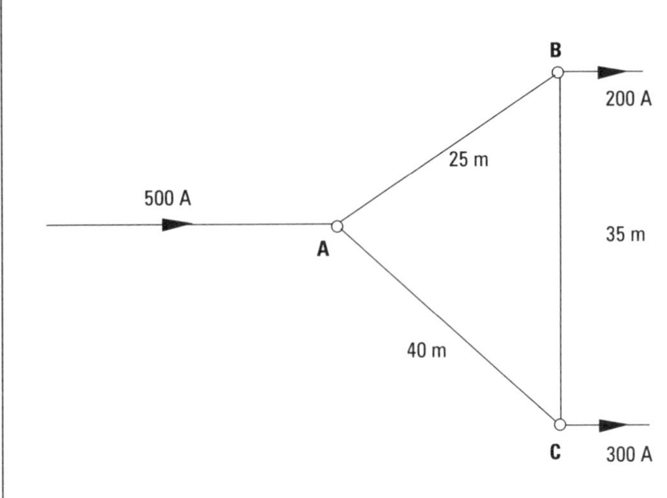

Figure 3.14 *A ring distribution system*

This is a Kirchhoff's Law problem and we'll remind ourselves of the basic facts:

1. The sum of the currents at every junction is ZERO.
2. Because it is a closed loop, the sum of the voltage drops in the loop is ZERO. Draw the loop again and put in the bits that we know.

Assume the current between **A** and **B** to be *X* Amperes and then apply Kirchhoff's Law to every junction from there on.

Then there comes the problem of calculating the voltage drops, because remember we have to add these up and the result must be ZERO.

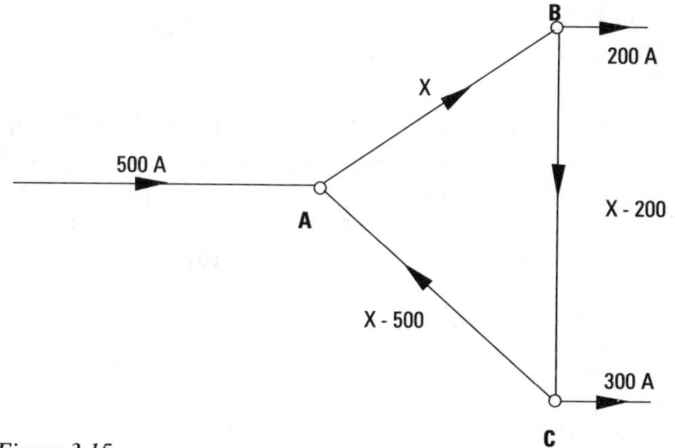

Figure 3.15

Currents

A to **B** $= X$ amps
B to **C** $= (X - 200)$ amps
C to **A** $= (X - 500)$ amps

Resistances at 0.0006 ohm/m

A to **B** $= 25 \times 0.0006 = 0.015\ \Omega$
B to **C** $= 35 \times 0.0006 = 0.021\ \Omega$
C to **A** $= 40 \times 0.0006 = 0.024\ \Omega$

The sum of the voltage drops can now be expressed as:

$$(I_{AB} \times R_{AB}) + (I_{BC} \times R_{BC}) + (I_{CA} \times R_{CA}) = 0$$

$(X \times 0.015) + ((X - 200) \times 0.021) + ((X - 500) \times 0.024) = 0$
(multiply out the brackets)

$0.015X + 0.021X - 4.2 + 0.024X - 12 = 0$
(take numbers to right hand side)

$0.015X + 0.021X + 0.024X = 16.2$

$0.06X = 16.2$

$X = 270$ A

The currents are:

A to **B** $= X = 270$ A

B to **C** $= (X - 200) = 70$ A

C to **A** $= (X - 500) = -230$ A (i.e. 230 A in the opposite direction)

Now prove the result by calculating the volts drop in the whole loop.

$(270 \times 0.015) + (70 \times 0.021) + (-230 \times 0.024)$

$4.05 + 1.47 - 5.52 = 0$

The solution is correct.

Try this

The circuit shown in the diagram is all connected with cable having a resistance of 0.124 mΩ/m.

Calculate the current in each branch of the circuit.

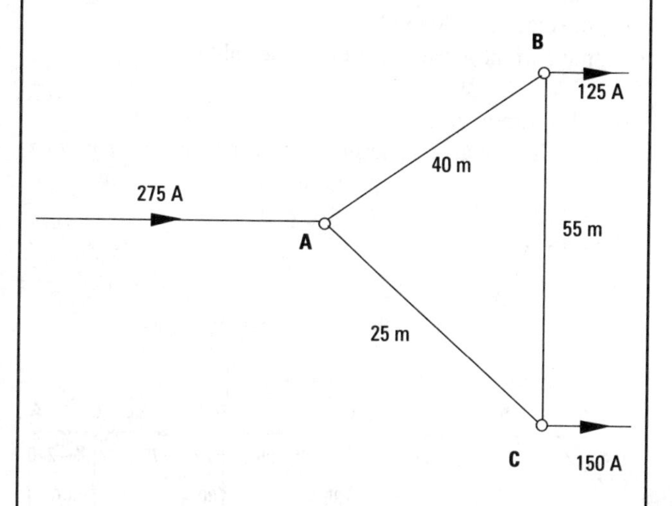

Now take a look at a very similar example.

A ring main distributor **ABCDEA** is fed at A with 240 A. The loads at **B D** and **E** are 60A, 80A and 65A respectively. The cable is of uniform cross-section and the lengths are as follows; **AB** 40m, **BC** 60 m, **CD** 35 m, **DE** 65 m, **EA** 25 m.

Determine
(a) the current in load **C**
(b) the current in each section of the cable

(a) The current at **C**
= Current at **A** – the sum of the currents at the other junctions
= 240 – 205
= 35 A

(b)

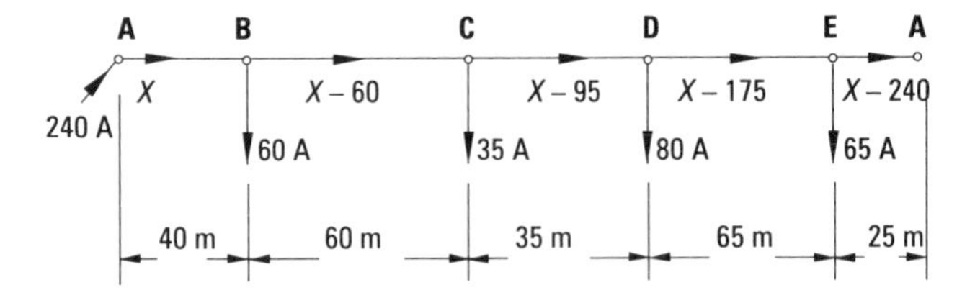

Figure 3.16

Note: The conductor resistance is not given, but since the cross sectional area is the same throughout, the conductor length can be used as an equivalent quantity.

If the current between **A** and **B** is X Amps then the others are as follows:

BC	$= (X - 60)$ amps
CD	$= (X - 95)$ amps
DE	$= (X - 175)$ amps
EA	$= (X - 240)$ amps

The sum of the equivalent voltage drops is therefore:

$$40X + 60(X-60) + 35(X-95) + 65(X-175) + 25(X-240) = 0$$

giving

$$40X + 60X - 3600 + 35X - 3325 + 65X - 11375 + 25X - 6000 = 0$$

Collecting the X terms on the left-hand side and the numbers on the right-hand side gives:

$$225X = 24\,300$$
$$X = 108$$

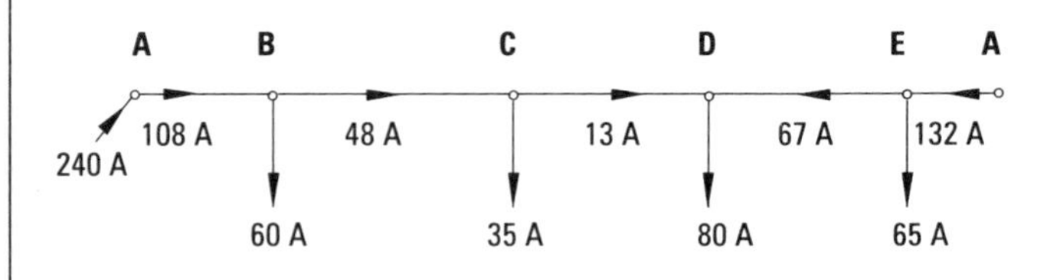

Figure 3.17

The currents are therefore:

A – B =	108 A
B – C =	48 A
C – D =	13 A
E – D =	67 A
A – E =	132 A

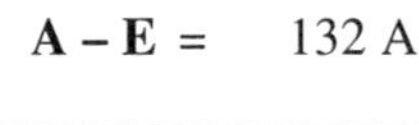

Try this

A two-core ring distributor of uniform cross-sectional area is loaded as shown in the diagram. The distributor is supplied at **A** with 240 V and the combined conductor resistance (supply and return) is 0.13 Ω per hundred metres.

Calculate:
(a) the current in each section of the distributor
(b) the voltage drop from **A** to **C**
(c) the power loss in section **B C**.

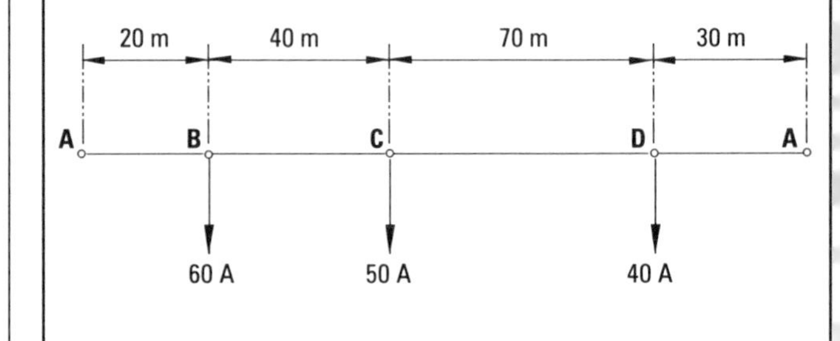

Bridge circuits

The balanced bridge circuit is a device frequently used for measurement purposes in electrical work. The commonest form is attributed to Charles Wheatstone, a 19th century inventor, a man of considerable talent who invented many things including the concertina and the stereoscope (a kind of 3D slide viewer).

The Wheatstone Bridge

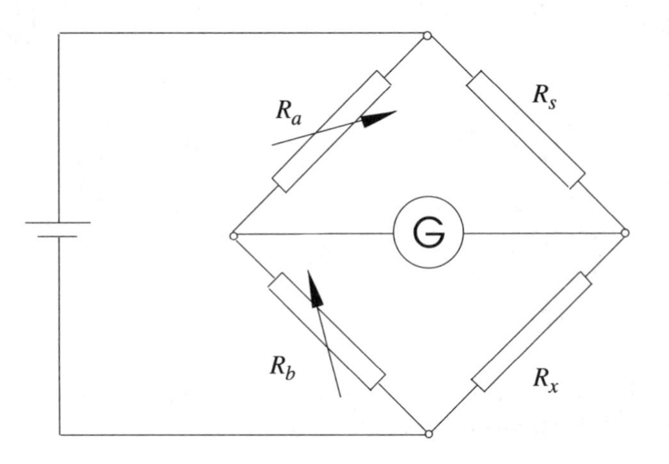

Figure 3.18 *Basic Wheatstone bridge circuit*

The device with the symbol G is a galvanometer, a very sensitive current indicating device which is used to detect the NULL condition, i.e. when no current is flowing through the instrument.

The resistors R_a and R_b form the " ratio arms" R_s is a standard resistor of known value and R_x is the unknown resistor, the value of which is to be measured.

Resistors R_a and R_b are adjusted until a NULL condition is obtained and at this stage the bridge is said to be "balanced".

At balance: $\frac{R_a}{R_b} = \frac{R_s}{R_x}$

and by transposition the value of

$$R_x = \frac{(R_s \times R_b)}{R_a}$$

Example

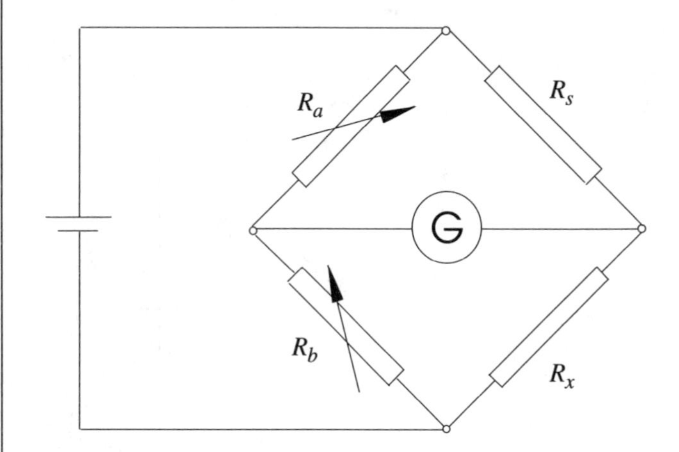

Figure 3.19

Solve for R_x:

R_a = 1000 Ω
R_b = 447 Ω
R_s = 250 Ω
R_x = ?

$$R_x = \frac{(R_s \times R_b)}{R_a}$$

$$R_x = \frac{250 \times 447}{1000}$$

$$= 111.75\ \Omega$$

The value of the unknown resistor is 111.75 Ω.

Practical uses for the balanced bridge circuit

The Murray Loop Test

If a multi-core underground cable develops an earth fault on one of its cores the location of fault has to be established so that the cable can be dug up and repaired.

It would be expensive and impractical to dig up any great length of cable so a fault location method, based on the balanced bridge can be used to direct the repair team to the position of the fault.

To conduct a Murray Loop Test, there must always be at least one sound core of the same cross-section as the faulty one.

The test is based on the Wheatstone Bridge principle and requires two, accurately calibrated variable resistors to form the ratio arms and a galvanometer or similar null indicator.

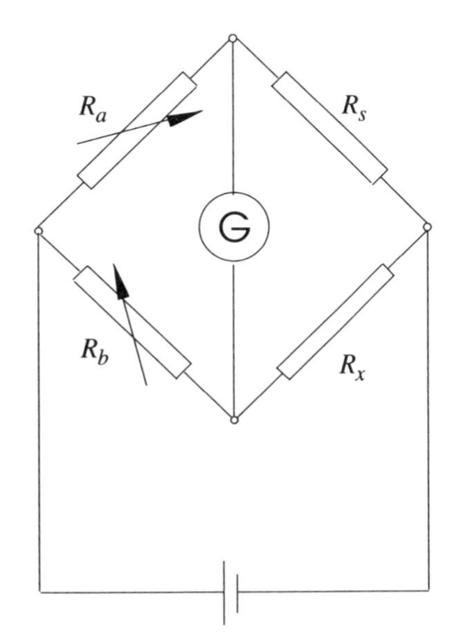

Figure 3.20 Wheatstone Bridge

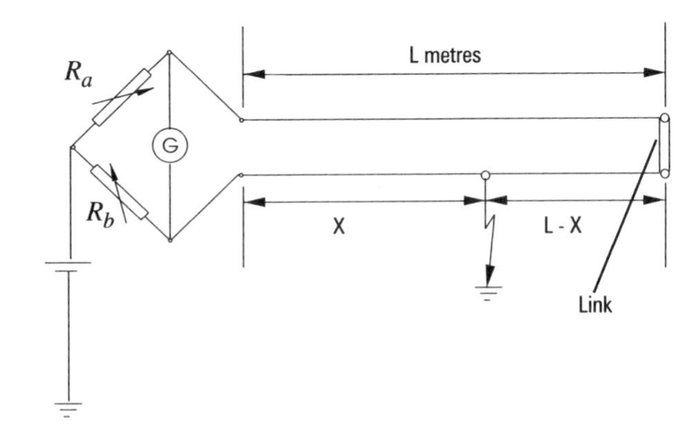

Figure 3.21 Murray Loop Test

In a practical situation, the cable would be exposed at both ends and the cores tested individually with an insulation resistance tester in order to identify the sound and faulty cores.

Once these have been identified, a shorting link is placed between them at the remote end.

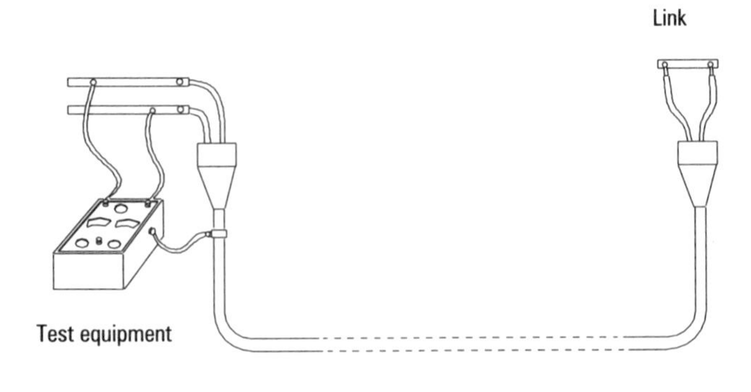

Figure 3.22 The circuit set up in practice

At the test end, the test gear is connected to earth and the operator adjusts the ratio arm resistors until a null balance is obtained and from the result, the location of the fault can then be calculated from the readings.

It works like this:

At balance, the sum of the two ratio resistors is proportional to the resistance of the whole conductor length, go and return. i.e. twice the cable length or $2L$ (metres).

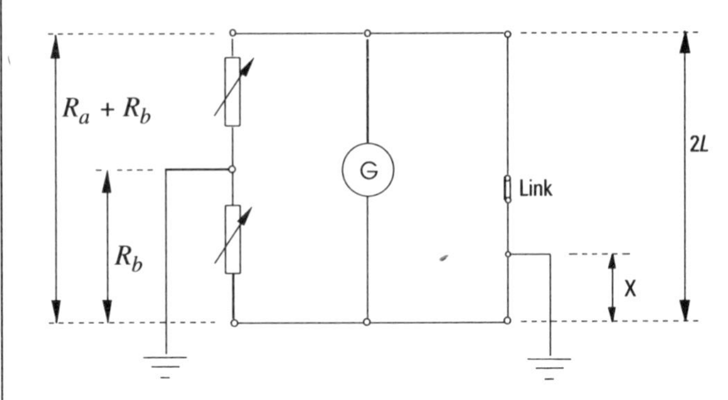

Figure 3.23

The lower of the two ratio resistors is proportional to the distance from the fault X (metres).

This gives us an expression based on this proportionality.

$$\frac{R_b}{(R_a + R_b)} = \frac{X}{2L}$$

which transposed becomes

$$X = \frac{2L \times R_b}{(R_a + R_b)}$$

In operator's terms:

"Twice the length times the part over the whole".

Example:

A twin cable 250 metres long develops an earth fault on one of its cores. A Murray Loop Test is set up and when balanced, the ratio resistors give a reading of $R_a = 600\Omega$ and $R_b = 122\Omega$

What is the distance to the fault?

"Twice the length times the part over the whole"

$$X = \frac{2L \times R_b}{(R_a + R_b)}$$

$$X = \frac{500 \times 122}{722}$$

$X = 84.48$ metres (from the test end)

Try this

A twin cable 1600 metres long develops an earth fault in one of its cores. A Murray Loop Test balances at $R_a = 450\ \Omega$ and $R_b = 394\ \Omega$

Determine the distance from the test end to the fault.

The Slide-wire Bridge

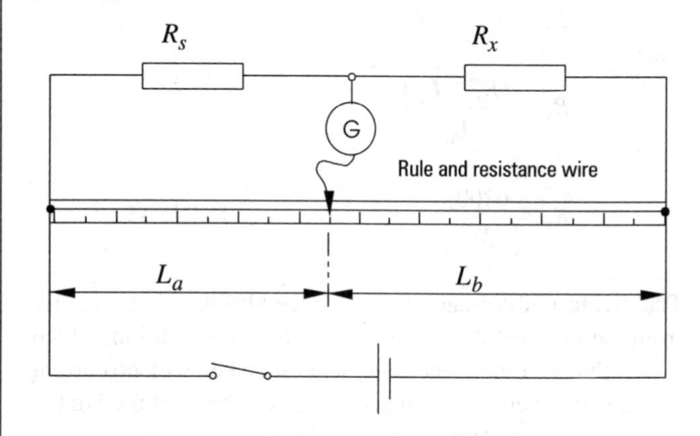

Figure 3.24 Slide wire bridge with 1 m rule and resistance wire.

This very basic instrument can be constructed from simple components and used to measure resistance to a reasonable degree of accuracy.

A one metre length of uniform cross-section resistance wire is fixed to a one metre wooden rule.

A sliding contact can be placed on the wire at any point and the length of wire either side of the contact represents the ratio arms. There is no need to know the actual resistance as this will be proportional to length because of the uniformity of the wire.

Example

The slide-wire bridge balances with the sliding contact 62 cm from the lower end of the scale and $R_s = 100\ \Omega$.

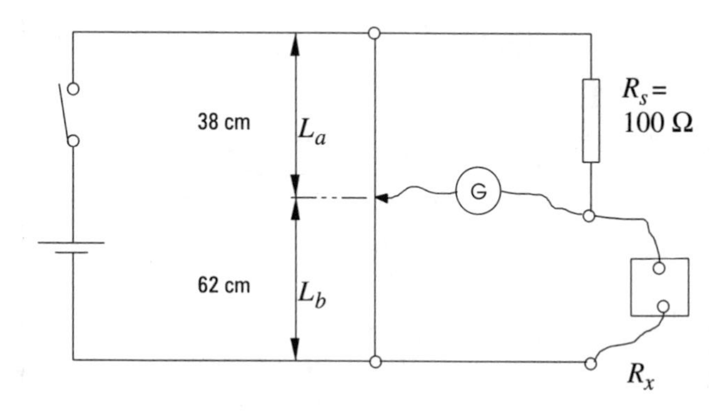

Figure 3.25 Circuit of slide wire bridge

The unknown resistance is found by

$$\frac{L_a}{L_b} = \frac{R_s}{R_x}$$

$$R_x = \frac{(R_s \times L_b)}{L_a}$$

$$R_x = \frac{6200}{38} \quad = \quad 163.16\ \Omega$$

The greatest advantage of the balanced bridge circuit comes from the fact that it is balanced in a NULL condition, which means that the crucial measurements are taken with no current flowing and therefore no inaccuracies can be attributed to the presence of a test current in the indicating instrument.

Try this

The slide-wire bridge balances as shown in the diagram. Calculate the value of R_x.

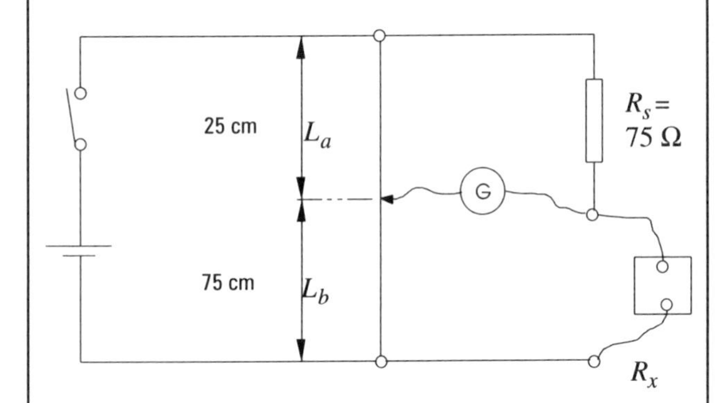

Alternating current circuits

When dealing with problems in a.c. circuits we will be using r.m.s. quantities wherever possible so that, as far as Ohm's Law is concerned, the calculations will be very similar to d.c. Similar to, but not exactly the same as d.c., because now we must consider impedance rather than resistance and also the effects of "phase angle" when various quantities come together when the circuit contains inductive or capacitive components.

Resistance

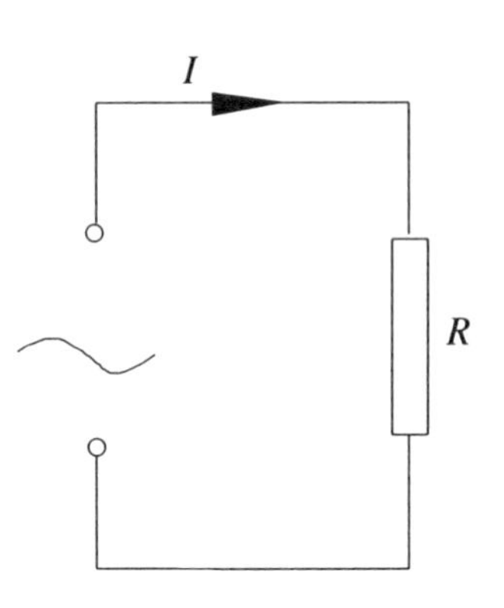

Figure 3.26 Resistive circuit

In a purely resistive circuit,

$$R = \frac{U}{I}$$

$$I = \frac{U}{R}$$

$$U = I \times R$$

Reactance

In practical terms, the reactance of a purely reactive component can be found by:

For an inductor, $X_L = 2\pi f L$

For a capacitor $X_C = \dfrac{1}{2\pi f L}$

and the value is expressed in ohms.

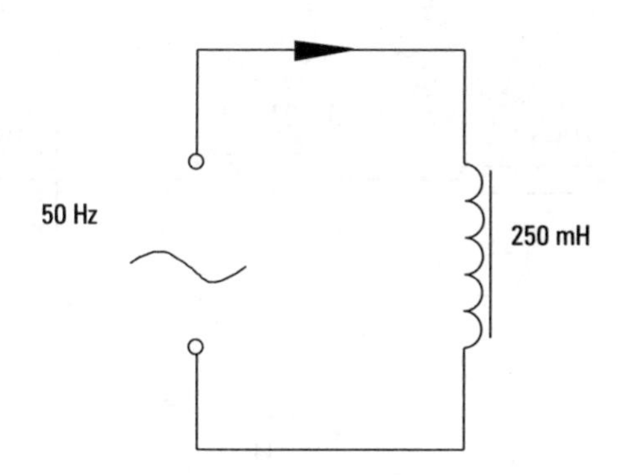

Figure 3.27 *Induction circuit*

(1) $X_L = 2\pi f L$

$= 2 \times \pi \times 50 \times 0.25$

$= 78.54\ \Omega$

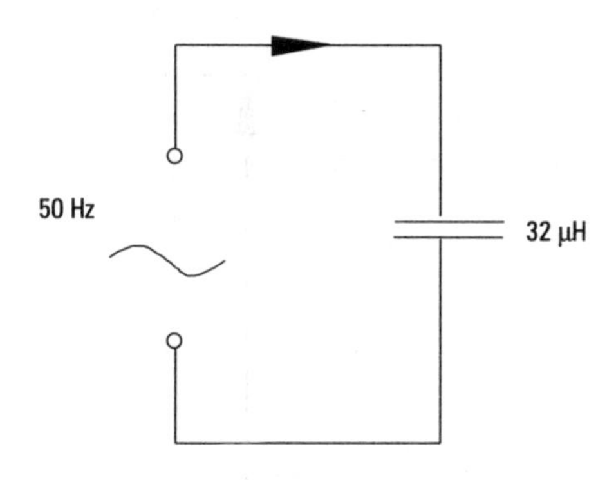

Figure 3.28 *Capacitive circuit*

(2) $X_C = \dfrac{1}{2\pi f C}$

$= \dfrac{1}{2 \times \pi \times 50 \times 32 \times 10^{-6}}$

$= 99.47\ \Omega$

Impedance

When a pure resistance and a pure reactance are combined in the same circuit, their effects are combined to produce a third quantity which is the **IMPEDANCE** of the component.

It is the impedance which restricts the flow of current in the circuit and since this quantity has the symbol Z, care must be taken to ensure that Ohm's Law is always used in the form:

$I = \dfrac{U}{Z}$ and its derivatives.

This must be used for all a.c. circuits even in the case of pure resistance since resistance and impedance are one and the same thing in these circumstances.

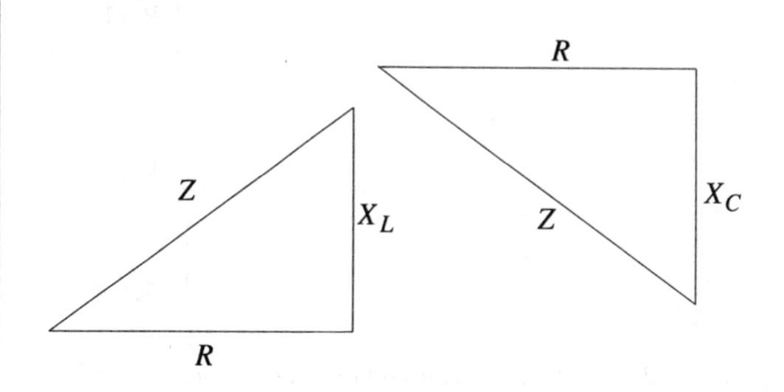

Figure 3.29

In graphical form, the impedance can be shown as a right-angled triangle as in Figure 3.29 above.

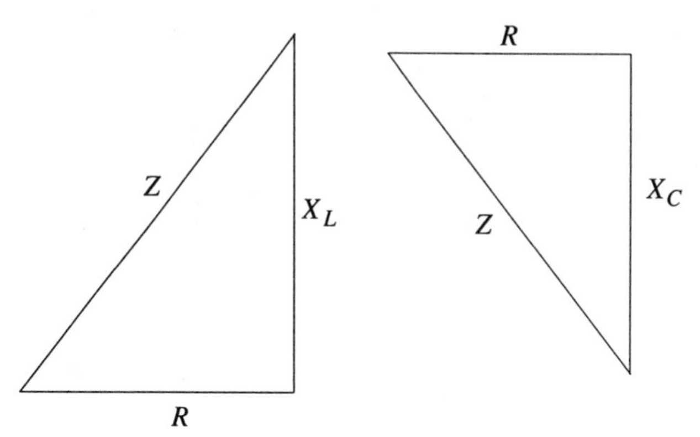

Figure 3.30

$R = 12\ \Omega$

$X_L = 16\ \Omega$

$Z = 20\ \Omega$

$R = 15\ \Omega$

$X_C = 20\ \Omega$

$Z = 25\ \Omega$

By calculation; the theorem of Pythagoras can be used and the result obtained by:

$Z = \sqrt{R^2 + X_L{}^2}$

$Z = \sqrt{144 + 256}$

$Z = 20\ \Omega$

$Z = \sqrt{R^2 + X_C{}^2}$

$Z = \sqrt{225 + 400}$

$Z = 25\ \Omega$

When a circuit contains resistance, inductance and capacitance, all three can be combined as in Figure 3.31.

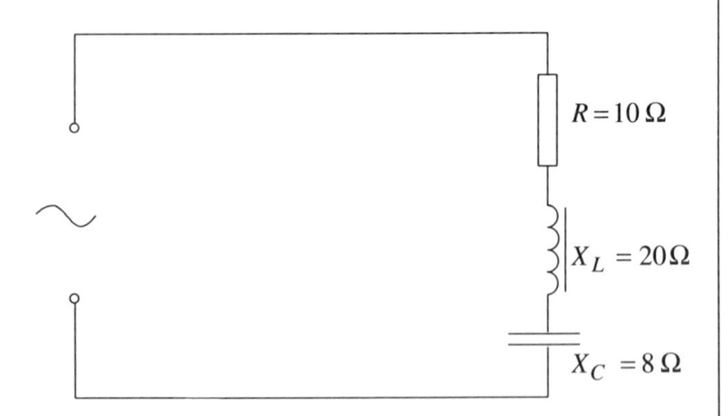

Figure 3.31 *Series circuit with resistance, inductance and capacitance*

The combined reactance is the difference between X_L and X_C.

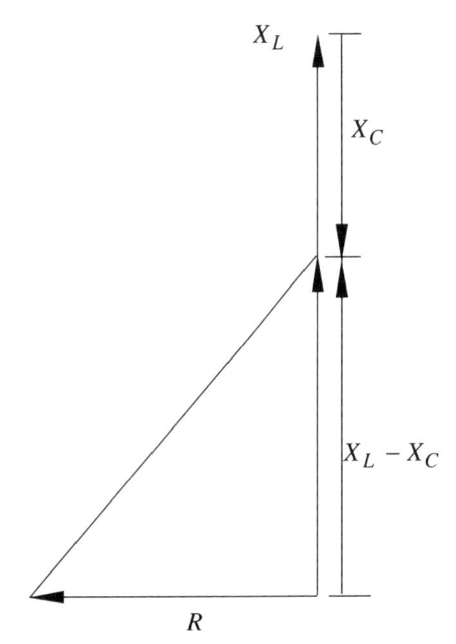

Figure 3.32 *Phasor diagram of series circuit*

$R = 10\ \Omega$ $X_L = 20\ \Omega$ $X_C = 8\ \Omega$

And by calculation this becomes

$$Z = \sqrt{(R^2 + (X_L - X_C)^2)}$$

$$Z = \sqrt{(10^2 + (20 - 8)^2)}$$

$$Z = \sqrt{244}$$

$$= 15.62\ \Omega$$

Example:

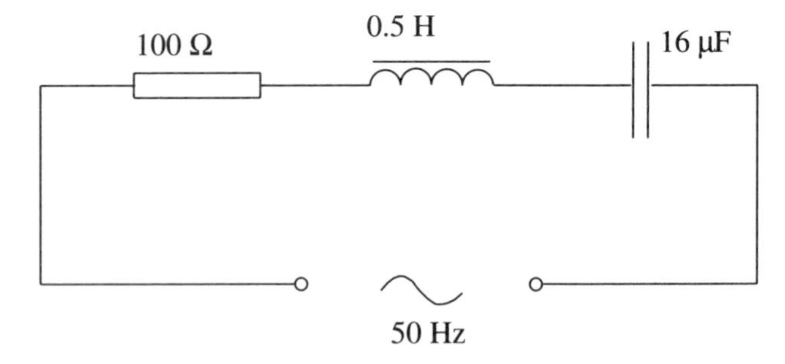

Figure 3.33

$X_L = 157\ \Omega$

$X_C = 200\ \Omega$ (very nearly) and

$R = 100\ \Omega$

Graphically:

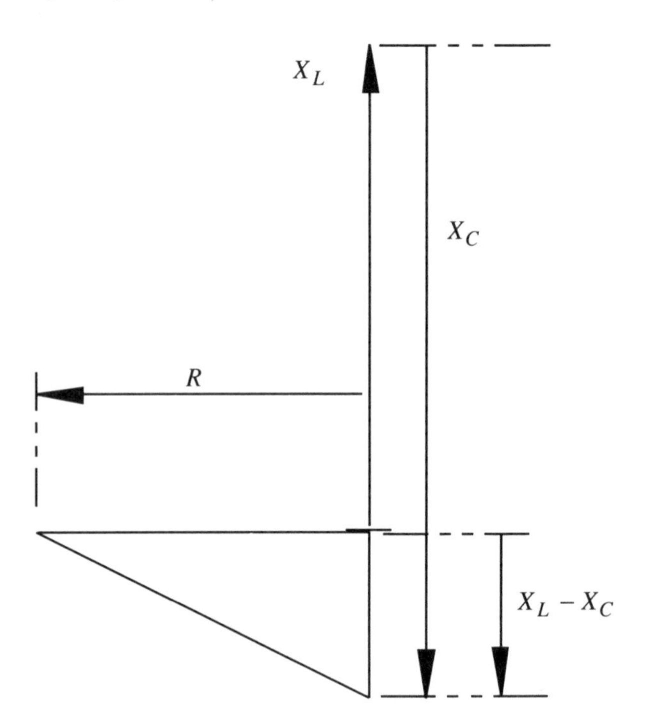

Figure 3.34

By calculation:

$$Z = \sqrt{(100^2 + (157 - 200)^2)}$$

$$Z = 108.85\ \Omega$$

Try this

For the circuit shown determine the

(a) inductive reactance

(b) capacitive reactance

(c) impedance

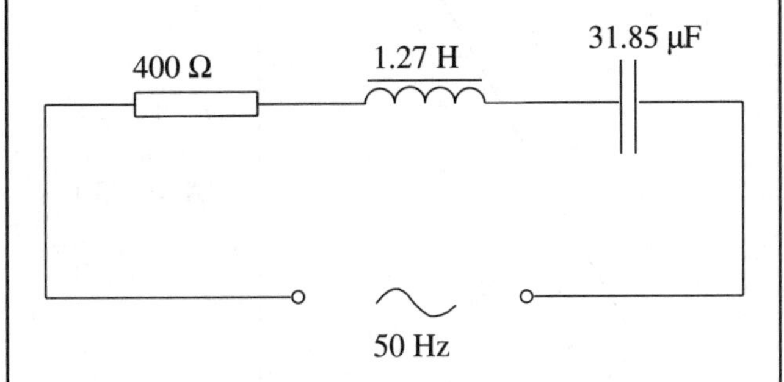

Phase angle, power and power factor

You may well have learned from previous studies that the reactive components in an a.c. circuit can cause the current to lag or lead the voltage by up to 90°.

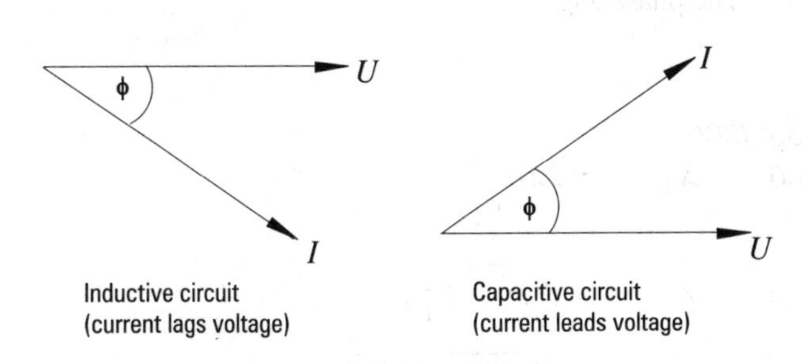

Figure 3.35

We are seldom if ever, asked to concern ourselves with the actual angle of lag or lead but we are frequently required to do something about its effects.

In a d.c. circuit, or an a.c. circuit which consists of pure resistance only, the power delivered in Watts is the product of the current and the voltage of the circuit concerned.

If however the circuit contains some reactive component, inductive or capacitive, then the current drawn from the supply is likely to be greater than that required for the power consumption of the load.

This effect is given the term "Power Factor" and is normally described as the ratio of Real Power, as may be measured in Watts, to Apparent Power which is merely the product of current and voltage.

There are three ways in which the power factor may be found:

1. Measure the power in Watts and divide by the product of the current and the voltage (the VoltAmps).
2. Divide the resistance of the circuit by its impedance.
3. Find the angle of lag or lead and determine its cosine.

Power factor is a ratio and is not a unit in its own right. Consequently it is expressed as a number between 0 and 1 with no units.

For a symbol it is usually given $\cos \phi$ (Φ is the upper case symbol), this being the cosine of the phase angle but may just as frequently be given the abbreviation pf instead.

Since $\cos \phi$ is derived from Watts, the appropriate expression for single phase power is:

$$P = UI \cos \phi$$

Example

A circuit having a resistance of 100 Ω and an inductance of 0.551 H is connected across a 200 V, 50 Hz a.c. supply.

Determine

(a) The reactance
(b) The impedance
(c) The current drawn from the supply
(d) The power factor
(e) The real power
(f) The phase angle

Solution

(a) $X_L = 2\pi f L$
$= 173.1\ \Omega$

(b) $Z = \sqrt{(R^2 + X_L^2)}$

$= \sqrt{10\,000 + 30\,000}$

$= 200\ \Omega$

(c) $I = \frac{U}{Z}$
$= \frac{200}{200}$
$= 1\ \text{A}$

(d) $\cos\phi = \frac{R}{Z}$
$= \frac{100}{200}$
$= 0.5$

(e) $P = U \times I \times \cos\phi$
$= 200 \times 1 \times 0.5$
$= 100\ \text{W}$

(f) By measurement

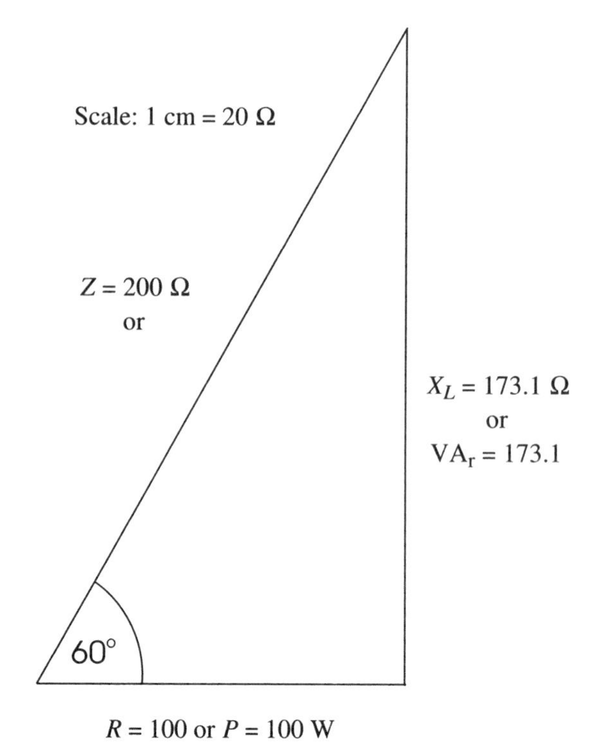

Figure 3.36 Impedance or power triangle

Closer examination of this example reveals that the power factor of 0.5 is also the cosine of 60° and that the power in Watts divided by the VoltAmps produces the same result.

Ohm's Law in its many forms gives us

$P = I^2 R$

$= 100\ \text{W}$

since the whole of the circuit current passes through the resistor.

Try this

A series circuit consisting of a resistor of 10 Ω and inductor of 63.7 mH and a capacitor of 100 μF is connected to a 50 V, 50 Hz supply. Determine:

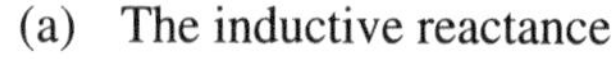

(a) The inductive reactance
(b) The capacitive reactance
(c) The impedance
(d) The circuit current
(e) The circuit power
(f) The power factor
(g) The phase angle between current and supply voltage
(h) The voltage measured across each component separately.

Phasor diagrams

The phasor diagram is a very useful tool and in the majority of cases can give a quick and reasonably accurate solution to an a.c. problem without having to resort to the calculator or memorise long and complicated formulae.

The phasor is a line, the length of which represents the magnitude (which is a posh way of saying the size) of a quantity and the angle at which it is drawn represents the phase angle.

We need phasor diagrams with a.c. because it is just not possible to add two and two to get four unless the twos are in phase with each other.

To perform a phasor addition of two quantities, draw one phasor to some suitable scale at its phase angle and where the first one finishes start the second, to the same scale and at its phase angle. The phasor sum is the distance between the start of the first phasor and the end of the other measured on the same scale.

Find the sum of two alternating currents.

I_1 = 30 A at 0°
I_2 = 45 A at 30° leading

Taking a scale of 2 cm = 10 A we get:

$I_1 + I_2$

I_2

30°

I_1

Figure 3.37

Note; The phase angles start from 0° which is horizontal and going to the right. The direction of rotation is anti-clockwise.

From this diagram we can see that the resultant current is very nearly 72.5 amperes.

Now let's try adding three currents.

I_1 = 10 A at 45°
I_2 = 10 A at 0°
I_3 = 15 A at 300°*

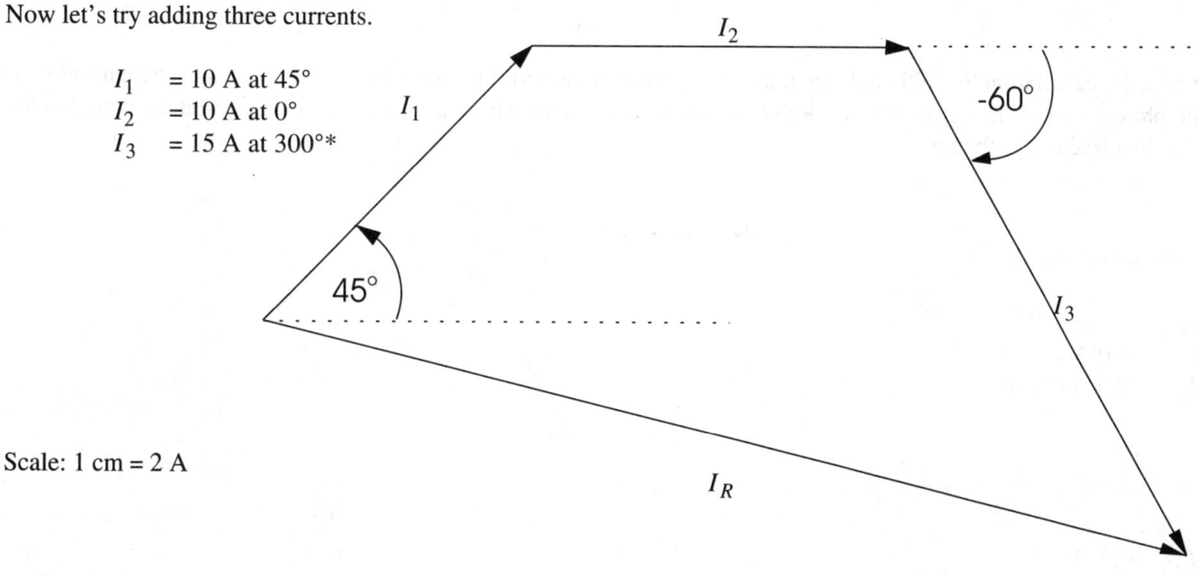

Scale: 1 cm = 2 A

Figure 3.37

***Note:** 300° in an anticlockwise direction is equivalent to minus 60°, i.e. 300° leading = 60° lagging.

As you can see from the phasor diagram, the resultant current is just a fraction over 25 A.

Try this

By means of a scaled phasor diagram, determine the sum of the three alternating currents.

I_a = 200 A at 0°
I_b = 160 A at 45°
I_c = 180 A at 80°

Three-phase currents

One of the most important applications of phasor addition is that of determining the current in the neutral of an unbalanced three phase load.

This can be done by calculation but, although accurate, this is a laborious and time consuming exercise. Since the current in the neutral is the phasor sum of the currents in the line conductors, a quick sketch on a scrap of paper can produce a pretty close approximation in a fraction of the time.

Example

What is the neutral current if

I_1 = 40 A at 0°
I_2 = 60 A at 120°
I_3 = 50 A at 240°

Solution

I_n = 18 A

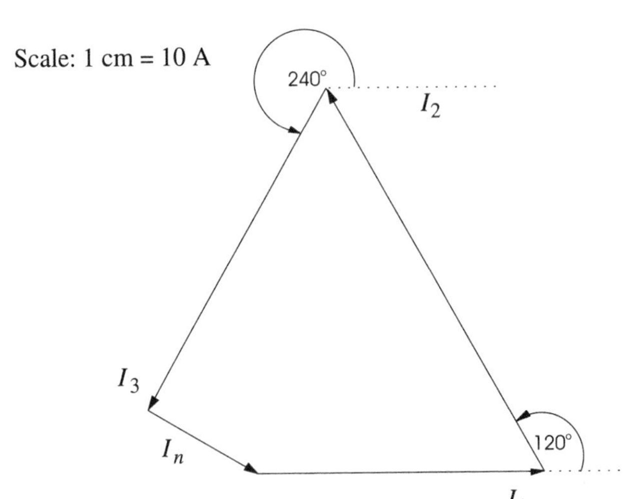

Figure 3.39

Three-phase circuits

A three-phase supply has three alternating voltages, each having the same amplitude and frequency but separated from each other by a phase angle of 120°.

When measured between phase and neutral each will display the same r.m.s. voltage but when observed simultaneously on an oscilloscope, the three traces will clearly show the phase difference between them.

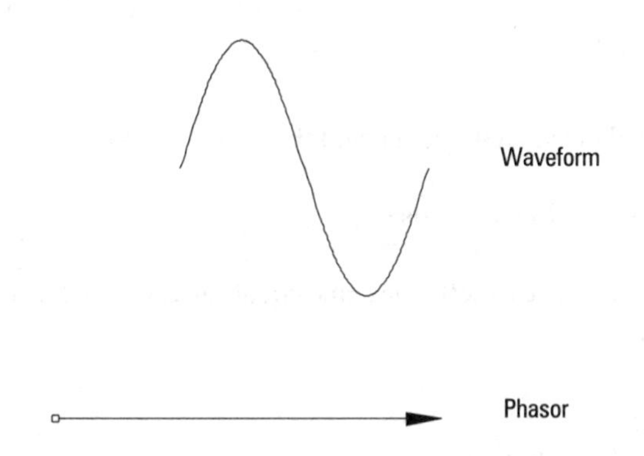

Figure 3.40 Single phase

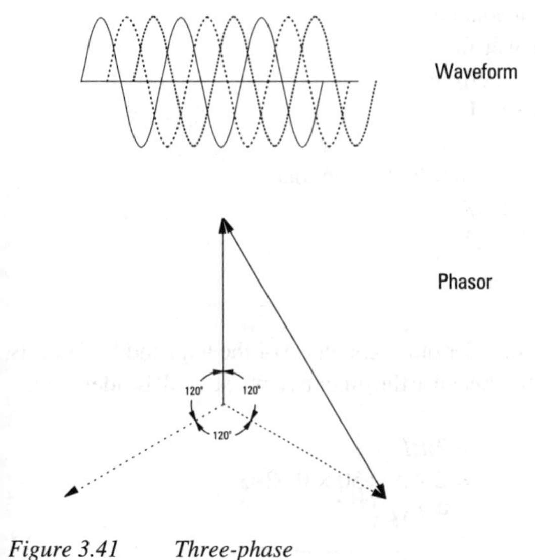

Figure 3.41 Three-phase

In the conventional low voltage system used in the UK mainland the r.m.s. voltage between phase and neutral is nominally 230 V.

Shown on a common phasor diagram the magnitude and phase displacement can be clearly seen.

The potential difference between any two phase voltages drawn to scale shows how the 400 V line voltage can be obtained.

Try this

Draw a scaled phasor diagram to represent a 220 V star connected supply. Measure the output of each phase to confirm the resultant delta voltages are 380 V.

Balanced loads in star and delta connection

Star

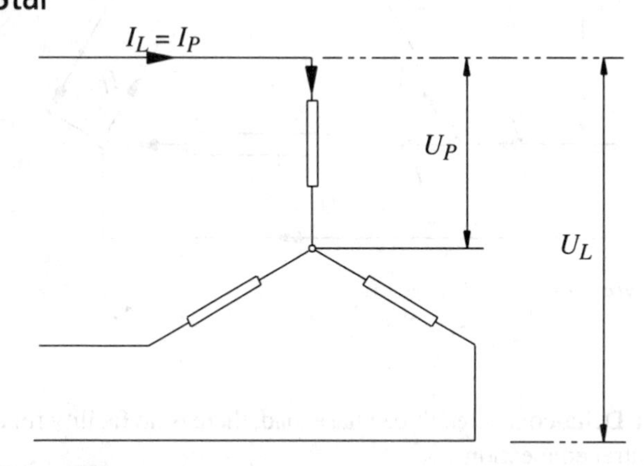

Figure 3.42

Three identical loads are connected as shown, with one end of each load connected to a different phase of the supply. The other ends are connected to a common (STAR) point which is conventionally connected to the neutral of the supply.

In this configuration, the neutral conductor carries the phasor sum of the three phase currents.

As you may well know, three equal currents each displaced from the other by an angle of 120° produce a phasor diagram which forms a perfect equilateral triangle with no resultant current. Hence the oft-quoted statement "there's no neutral current in a balanced three phase load".

If the loads are identical and the supply voltages are identical then why not consider a three phase balanced load to be three identical single phase loads.

Thus
Total Power = Phase Voltage × Phase Current × p.f. × 3

But the Phase Voltage is the Line Voltage divided by $\sqrt{3}$.

Total power = Line Voltage × Phase Current × p.f. × $\frac{3}{\sqrt{3}}$

But in a Star connected load:
line current = phase current
and $\frac{3}{\sqrt{3}} = \sqrt{3}$

So for a balanced three phase Star connected load

$$\text{Total Power} = \sqrt{3}\, U_L\, I_L \cos\phi$$

Delta

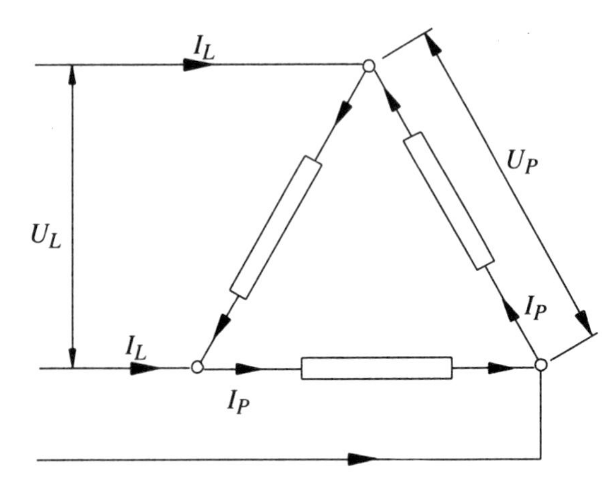

Figure 3.43

In a Delta-connected three phase load, there is no facility for a neutral connection.

In a balanced delta all three sections are identical but in each case each connection is made between two phases of the supply.

The total power delivered to the load is three times the power delivered to each phase of the load.

Thus;
Total power = Power in each load × 3
= Line voltage × phase current × cos ϕ × 3

But the phase current is the line current divided by $\sqrt{3}$

So the expression becomes

Total power = Line voltage × Line Current × cos ϕ × $\frac{3}{\sqrt{3}}$

but $\frac{3}{\sqrt{3}} = \sqrt{3}$

So for a balanced Delta-connected three phase load:

Total Power = $\sqrt{3}\, U_L\, I_L \cos\phi$

Which is of course exactly the same equation as for the STAR connection.

Example

A three phase balanced load consists of three identical windings each with a resistance of 6.9 Ω and an inductance of 29.3 mH. The supply voltage is 400 V and the frequency is 50 Hz.

Determine,
(a) The inductive reactance
(b) The impedance
(c) The power factor
(d) The line current:
  i) in STAR
  ii) in DELTA
(e) The power supplied to the load:
  i) in STAR
  ii) in DELTA

Solution

First of all consider only one phase of the load and find out its characteristics because the other two phases will be identical.

(a) $X_L = 2\pi fL$
$= 2 \times \pi \times 50 \times 0.0293$
$= 9.2\ \Omega$

(b) $Z = \sqrt{(R^2 + X_L{}^2)}$
$= \sqrt{132.25}$
$= 11.5\ \Omega$

(c) p.f. $= \frac{R}{Z}$
$= \frac{6.9}{11.5}$
$= 0.6$

(d) i) In Star connection, line current = phase current
phase voltage = 230 V

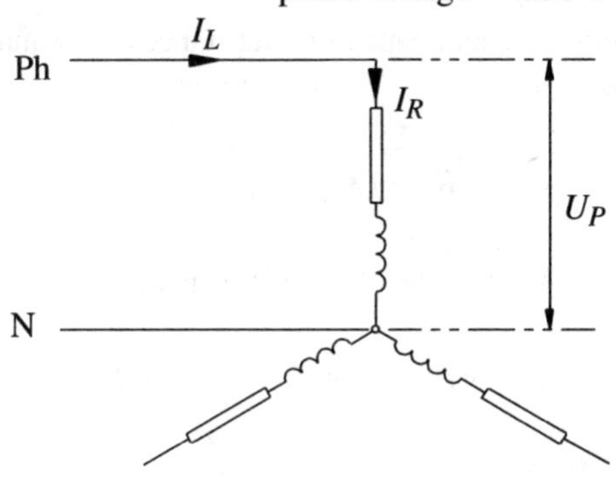

Figure 3.44

$$I = \frac{U}{Z}$$

$$= \frac{230}{11.5}$$

$$= 20 \text{ A}$$

(d) ii) In Delta connection,
line current = phase current $\times \sqrt{3}$
phase voltage = 400 V

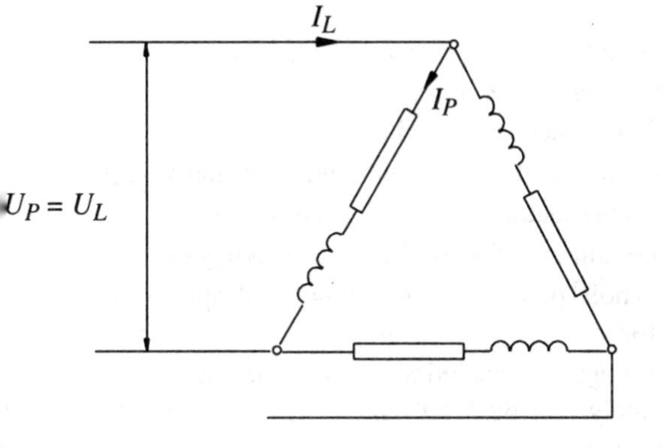

Figure 3.45

$$I_P = \frac{U}{Z}$$

$$= \frac{400}{11.5}$$

$$= 34.78 \text{ A}$$

$$I_L = I_P \times \sqrt{3}$$

$$= 60.24 \text{ A}$$

(e) i) In Star connection

$$P = \sqrt{3}\, U_L\, I_L \cos\phi$$

$$= 1.732 \times 400 \times 20 \times 0.6$$

$$= 8.313 \text{ kW}$$

(e) ii) In Delta connection,

$$P = \sqrt{3}\, U_L\, I_L \cos\phi$$

$$= 1.732 \times 400 \times 60.24 \times 0.6$$

$$= 25 \text{ kW}$$

> ***Remember***
> When frequency is 50 Hz then
>
> $$2\pi f L = 314$$
>
> 

Example

A 3 phase 400 V, 10 kW, 50 Hz, balanced, delta connected load has a lagging power factor of 0.707.

Determine the line current, the phase current, the impedance, resistance, reactance and inductance of the load.

(I know its a tall order but somebody might ask you, even an examiner.)

Since $P = \sqrt{3}\, U_L\, I_L \cos\phi$ then

$$I_L = \frac{P}{\sqrt{3}\, U_L \cos\phi}$$

$$I_L = \frac{10000}{\sqrt{3} \times 400 \times 0.707}$$

$$= 20.41 \text{ A}$$

$$I_P = \frac{I_L}{\sqrt{3}}$$

$$= 11.79 \text{ A}$$

$$Z = \frac{U_P}{I_P} \quad \text{(Delta connection } U_P = U_L\text{)}$$

$$= \frac{400}{11.79}$$

$$= 33.93\ \Omega$$

$$\cos\phi = \frac{R}{Z}$$

so $R = Z\cos\phi$

$$= 33.93 \times 0.707$$

$$= 24\ \Omega$$

$$X_L = \sqrt{(Z^2 - R^2)}$$

$$= 23.99\ \Omega$$

Since $X_L = 2\pi fL$ then

$$L = \frac{X_L}{2\pi f}$$

$$= 0.0764\ \text{H}\ (76.4\ \text{mH})$$

ALL VALUES ARE PER PHASE

It wasn't that bad after all.

Try this

A three phase load consists of three identical windings each having a resistance of 80 Ω and an inductive reactance of 60 Ω. Determine the current drawn from a 400 V three phase supply with the load connected

(a) in STAR

(b) in DELTA

Resonance

When the inductive and capacitive reactances are equal, the impedance of a series R L C circuit will be

$$Z = \sqrt{R^2 + (X_L - X_C)^2}$$

but since $X_L = X_C$ then the effect of this is that

$$Z = \sqrt{R^2} \quad \text{i.e.} \quad Z = R$$

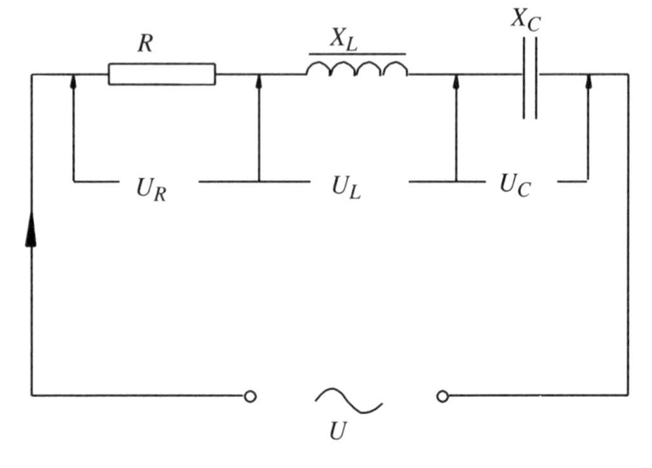

Figure 3.46

Under conditions of series resonance, the following circuit characteristics will be observed.

- The impedance will equal the resistance.
- The current will be at its maximum for that voltage.
- The power factor will be unity ($\cos \phi = 1$).
- The condition will exist for one frequency only.
- The whole of the supply voltage will appear across the resistor.
- The voltages measured across the capacitor and inductor separately will be $I \times X_L$ and $I \times X_C$ and can exceed the supply voltage.

Example

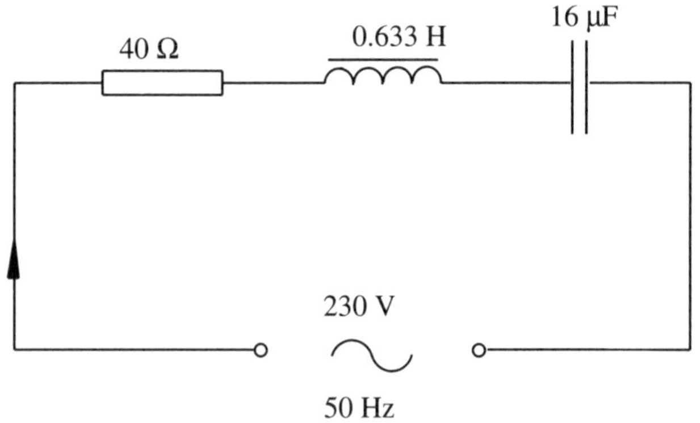

Figure 3.47

Taking the circuit in Figure 3.47:

$$R = 40\ \Omega$$
$$L = 0.633\ \text{H}$$
$$C = 16\ \mu\text{F}$$

$$X_L = 2\pi fL = 198.9\ \Omega$$

$$X_C = \frac{1}{2\pi fC} = 198.9\ \Omega$$

$$Z = 40\ \Omega$$

$$\cos\phi = \frac{R}{Z} = 1$$

$$I = \frac{U}{Z} = \frac{230}{40} = 5.75\ \text{A}$$

$$U_R = 40 \times 5.75 = 230\ \text{V}$$

$$U_L = 198.9 \times 5.75 = 1143.7\ \text{V}$$

$$U_C = 198.9 \times 5.75 = 1143.7\ \text{V}$$

This does not mean that Kirchhoff has gone mad but that the two voltages U_L and U_C are opposing each other by 180° therefore cancel each other out in terms of the complete circuit.

It is possible to determine the frequency which will cause resonance in a circuit if the capacitance and the inductance are known by applying this formula:

Resonant frequency

$$f_r = \frac{1}{2 \times \pi \sqrt{LC}}$$

Just try the last example

$$f_r = \frac{1}{2 \times \pi \sqrt{LC}}$$

$$= \frac{1}{2\pi \sqrt{(0.633 \times 16 \times 10^{-6})}}$$

$$= 50\ \text{Hz}$$

Example

What is the resonant frequency of the circuit shown in Figure 3.48?

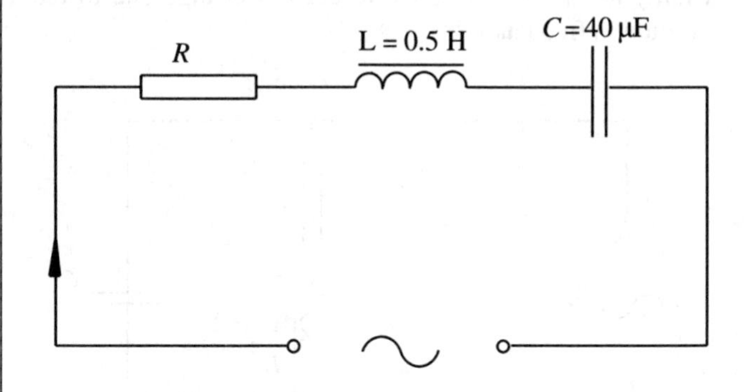

Figure 3.48

$$f_r = \frac{1}{2 \times \pi \sqrt{LC}}$$

$$= \frac{1}{2\pi \sqrt{(0.1 \times 40 \times 10^{-6})}}$$

$$= 79.57\ \text{Hz}$$

Try this

1. List FOUR characteristics of a series resonant circuit.

2. Calculate the resonant frequency of a series circuit consisting of a resistance of 24 Ω and an inductance of 50 mH and a capacitance of 100 μF.

Parallel resonance

When considering circuits such as the one in Figure 3.49 it is clear that at some particular frequency the power factor will be at unity as the effects of the reactances change due to the variation in frequency.

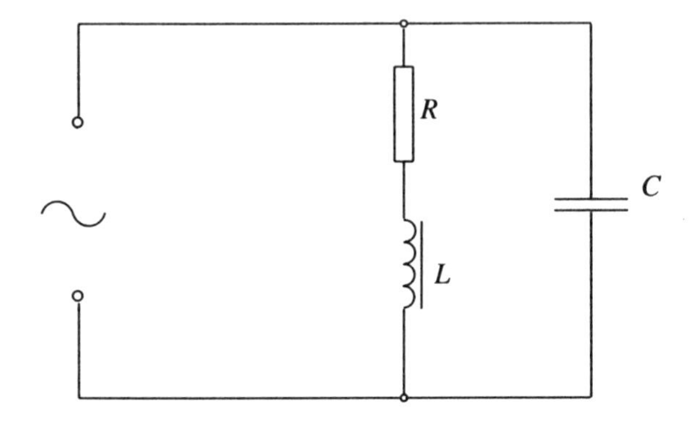

Figure 3.49

First consider the *RL* branch only.

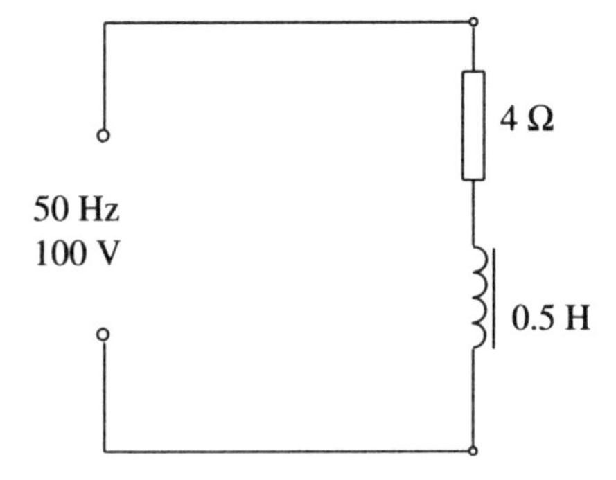

Figure 3.50

The inductive reactance at 50 Hz will be

$$X_L = 2\pi fL$$
$$= 2\pi f \times 0.5$$
$$= 157.08\ \Omega$$

and this in series with a 4 Ω resistor gives an impedance of

$$Z = \sqrt{R^2 + X_L^2}$$
$$= \sqrt{4^2 + 157.08^2}$$
$$= 157.13\ \Omega$$

So the current drawn from a 100 volt supply will be

$$I = \frac{U}{Z}$$

$$= \frac{100}{157.13} = 0.63464\ \text{A}$$

The power factor of this part of the load will be

$$\cos\phi = \frac{R}{Z} \text{ which is } 0.0255$$

which corresponds to an angle of lag of 88.54°.

Drawn as a phasor diagram:

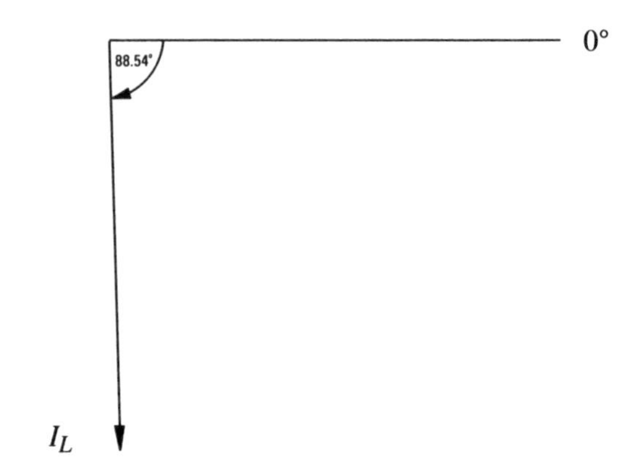

Figure 3.51

Now look at the capacitive branch.

The capacitor can be assumed to be almost pure reactance therefore

$$X_C = \frac{1}{2\pi fC}$$

$$= 159.15\ \Omega$$

100 V

$C = 20\ \mu\text{F}$

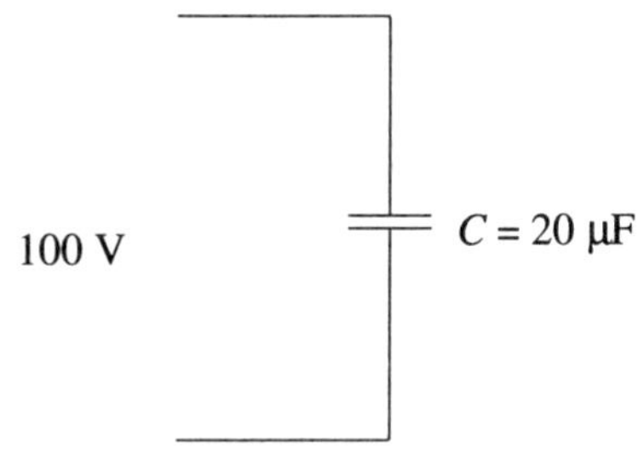

Figure 3.52

The current drawn by this component is going to be

$$I_C = \frac{U}{X_C}$$

$$= \frac{100}{159.15}$$

$$= 0.628\ \text{A}$$

which is very close to the current in the inductive branch but in this case leading by practically 90° as the inductive current was lagging by nearly the same angle.

Now add the two currents as a phasor diagram

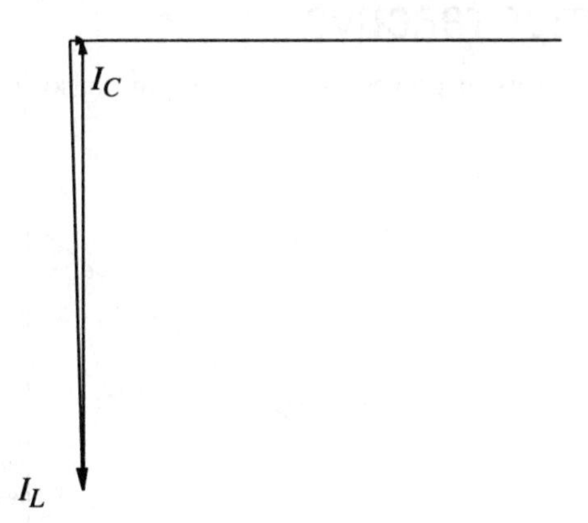

Figure 3.53

As you can see, the currents in the two branches are very large in comparison with the current drawn from the supply but this condition will only exist at resonant frequency

i.e. when

$$F = \frac{1}{2\pi\sqrt{LC}}$$

$$= \frac{1}{2\pi\sqrt{0.5 \times (20 \times 10^{-6})}}$$

$$= 50 \text{ Hz (or very nearly)}$$

To illustrate the point calculate the current at
(a) 25Hz and
(b) 100Hz

(a) $X_L = 2\pi \times 25 \times 0.5$
$= 78.5\ \Omega$

$Z = \sqrt{4^2 + 78.5^2}$
$= 78.64\ \Omega$

$\cos\phi = \frac{4}{78.64} = 0.05 \qquad \phi = 87°$

$I_L = \frac{100}{78.64}$

$= 1.275 \text{ A}$

$X_C = \frac{1}{2 \times \pi \times 25 \times 20 \times 10^{-6}}$

$= 318.47\ \Omega$

$I_C = \frac{100}{318.47}$

$= 0.314 \text{ A}$

The phasor sum becomes

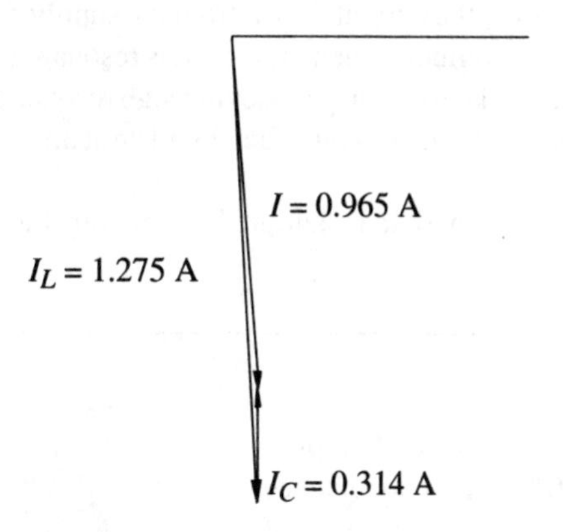

Figure 3.54

The current is more than at resonance and lagging.

(b) $X_L = 2 \times \pi \times 100 \times 0.5$

$= 314.16\ \Omega$

$Z = \sqrt{4^2 + 314.16^2}$

$= 314.18\ \Omega$

$\cos\phi = \frac{4}{314.18} = 0.0127 \qquad \phi = 89.27°$

$I_L = \frac{100}{314.18} = 0.318 \text{ A}$

$X_C = \frac{1}{2 \times \pi \times 100 \times 20 \times 10^{-6}} = 79.56\ \Omega$

$I_C = \frac{100}{79.56} = 1.256 \text{ A}$

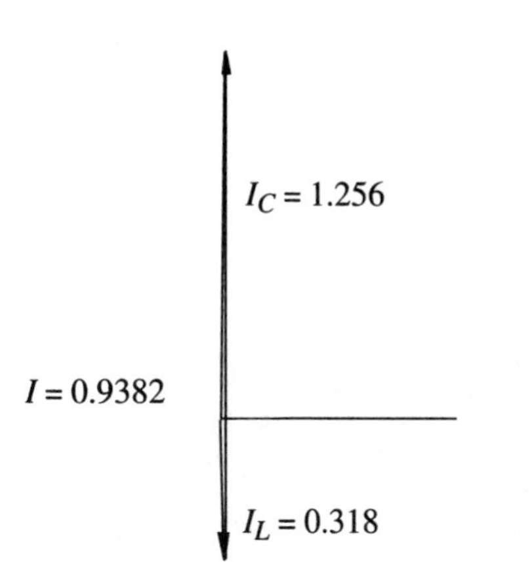

Figure 3.55 Phasor diagram

Again, the current is more than at resonance but in this case it is leading. From these two examples we may conclude that as the frequency is increased, the current drawn from the supply will reduce until it is at a minimum when the circuit is resonant and then increase again as the frequency passes through resonance. At resonant frequency only, the power factor will be at unity.

This is the principle which is adopted in power factor correction.

Try this

An a.c. circuit consists of a pure resistance of 10 Ω, a pure inductance of 250 mH and a pure capacitance of 120 μF all connected in parallel to a 100 V 55 Hz supply.

Determine the current in each branch and the current drawn from the supply.

VoltAmps, power, power factor and VoltAmps reactive

The components of power can be shown in phasor diagram form.

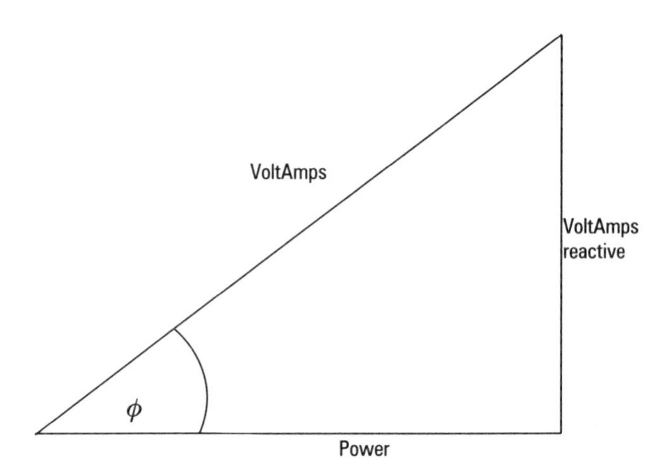

Figure 3.56

A great deal of information can be derived very quickly and simply from such a diagram. All that is necessary is a pencil, a ruler and a protractor. Of course a calculator will be necessary in order to obtain the angles involved.

Take a look at this circuit.

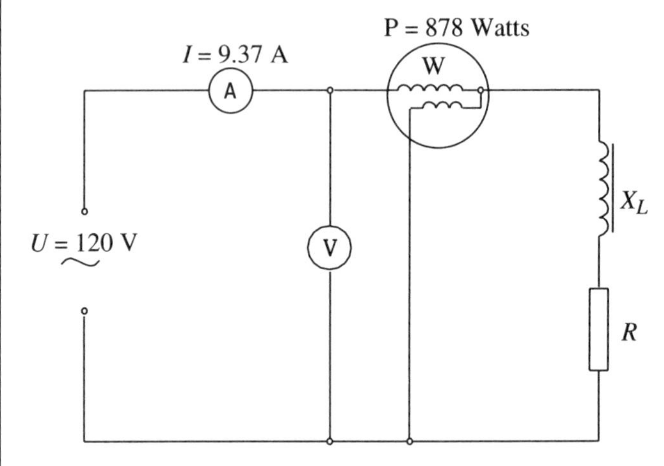

Figure 3.57

It consists of an inductor connected across a 120 V a.c. supply. An ammeter, voltmeter and wattmeter in the circuit will tell us all that we need to know.

From the readings taken we find that

$$U = 120 \text{ V}$$
$$I = 9.37 \text{ A}$$
$$P = 878 \text{ W}$$

We can calculate the VoltAmps from

$$VA = 120 \times 9.37$$
$$= 1124.4 \text{ VA}$$

From $P = UI\cos\phi$

we get $\cos\phi = \dfrac{\text{Watts}}{\text{VA}}$

$$= \frac{878}{1124.4}$$

$$= 0.78$$

Leave the 0.78 on the calculator display and follow the instructions for $\cos^{-1}$ which should produce the angle 38.74°.

Now draw the triangle:

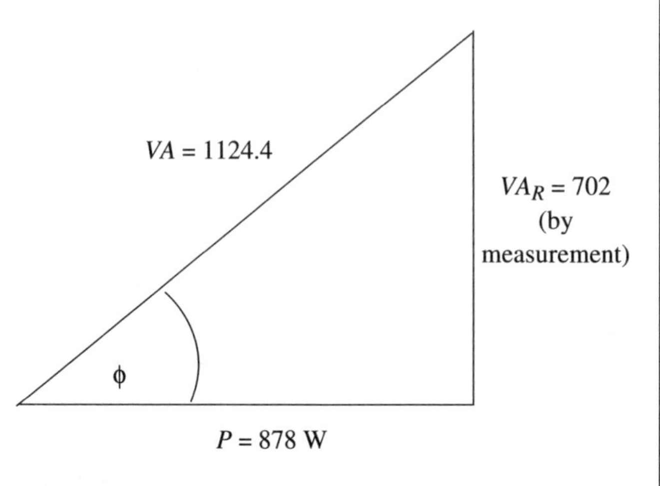

Figure 3.58

Now using your protractor to measure the angle, draw the triangle which should have a perfect right angle.

We should now have the following information
(a) the VoltAmps
(b) the power factor
(c) the phase angle and
(d) the VoltAmps reactive (by measuring the vertical side).

Alternatively:

You can do it without knowing the angle beforehand if you have a pair of compasses.

Draw a horizontal line representing the power in Watts.

From the start of this line draw an arc to the same scale, representing the VoltAmps.

From the end of the power phasor draw a vertical line to cut the arc.

Join the three points and you have your power triangle.

Try it now using the values from the previous example and you will have a triangle of the same dimensions with the angle as before.

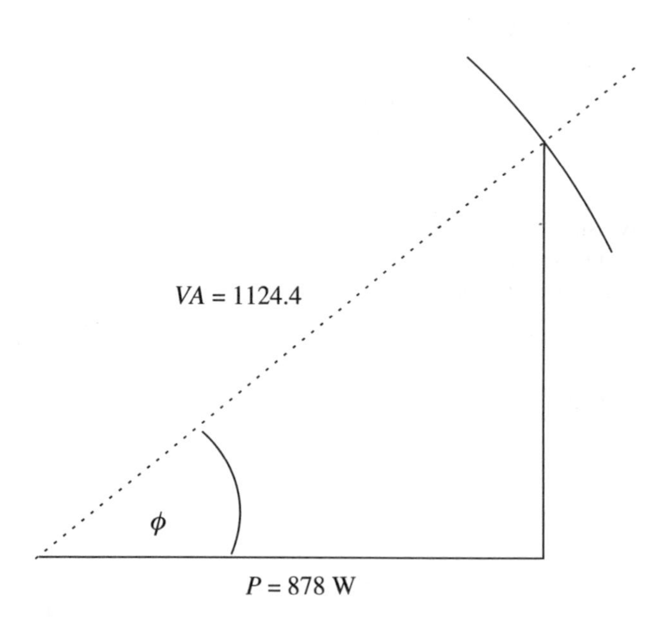

Figure 3.59

Trigonometrical (Trig.) ratios

Calculated values using the trigonometrical ratios sine, cosine and tangent may be more difficult but are extremely accurate.

$$\cos\phi = \frac{\text{Adj}}{\text{Hyp}} = \frac{\text{Watts}}{\text{VoltAmps}}$$

$$\sin\phi = \frac{\text{Opp}}{\text{Hyp}} = \frac{\text{VA}_r}{\text{VoltAmps}}$$

$$\tan\phi = \frac{\text{Opp}}{\text{Adj}} = \frac{\text{VA}_r}{\text{Watts}}$$

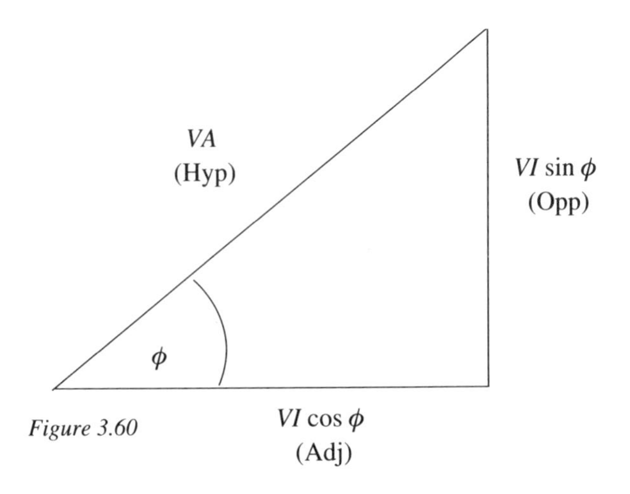

Figure 3.60

$VI \cos \phi$
(Adj)

By using these ratios it is always possible to find out all you need to know about a power (or impedance) triangle having been given at least one side and an angle or one of its Trig. ratios.

Example

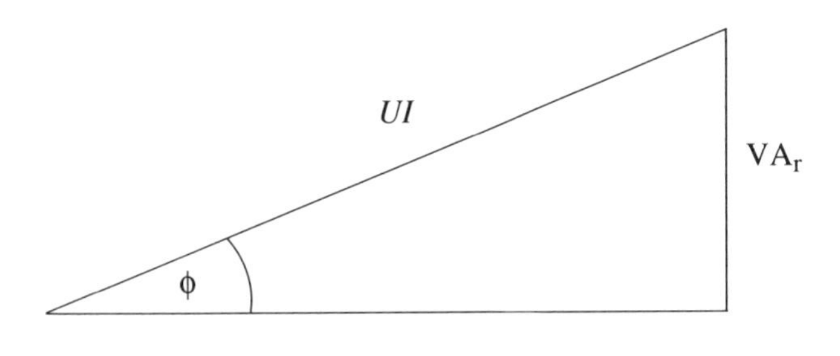

$UI \text{ Cos } \phi$

Figure 3.61

$U = 155$ V
$I = 13.47$ A
$\sin \phi = 0.37$

If $\sin \phi$ is 0.37,
ϕ = inv sin 0.37 (or $\sin^{-1}$)
= 21.715°

Then $\cos \phi = 0.93$ giving
$UI \cos \phi = 1939$ W and
$UI \sin \phi = 772.5 \text{ VA}_r$

There's not much else you would need to know.

When working on three phase balanced loads, a similar approach may be adopted.

Try this

A single phase 230 V load has a load current of 22 A and a power consumption of 4048 W.

Determine, by calculation:

(a) the VA_r of the load and
(b) the angle between the current and the supply voltage

Example

A three phase balanced load takes a line current of 18 A from a 400 V supply, a three phase wattmeter indicates a total power consumption of 10 kW. The load is DELTA connected and the current is lagging.

The VoltAmps of the load is
$\sqrt{3} \times 400 \times 18 = 12.470$ kVA

The total power is 10 kW.

Therefore the power factor is
$$\frac{10}{12.470} = 0.8$$

The phase angle is
inv cos 0.8 = 36.87°

The sine of this angle is 0.6.

The reactive component
$= UI \sin \phi$
= 7.482 kVAr

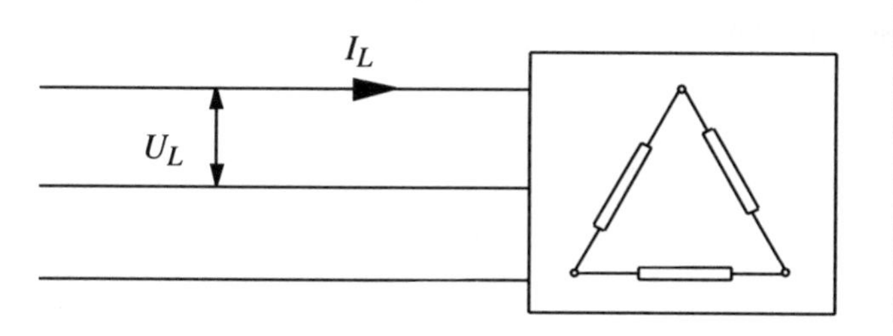

Figure 3.62

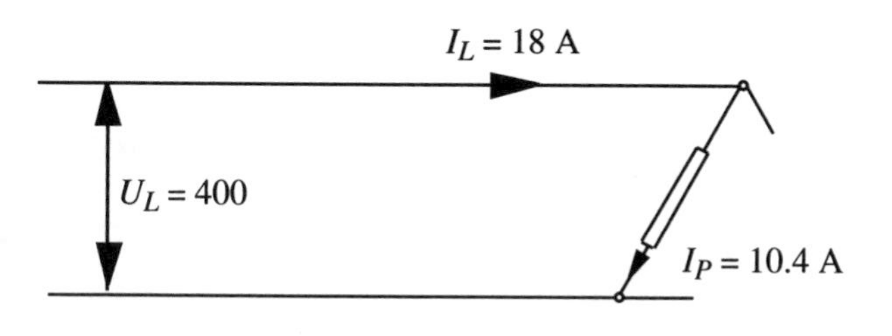

Figure 3.63

Consider one phase only, as the other two are identical and for a DELTA connection.

$$I_P = \frac{I_L}{\sqrt{3}}$$

$$= \frac{18}{1.73}$$

$$= 10.4 \text{ A}$$

$$\text{Therefore } Z = \frac{U_L}{U_P}$$

$$= \frac{400}{10.4}$$

$$= 38.46\ \Omega$$

The other two components R and X_L can be found from the impedance triangle or by calculation.

Exercises

1. a) Show by means of a circuit diagram how a balanced bridge circuit can be used to locate an earth fault on one core of a twin cable.
   b) Determine the position of an earth fault in a twin cable 44 metres long if the bridge balances at $R_a = 550\ \Omega$ and $R_b = 247\ \Omega$.

2. Determine, by phasor addition, the neutral current in an unbalanced star when
   $I_r = 45$ A leading 0° with a power factor of 0.8
   $I_y = 30$ A in phase at 240°
   $I_b = 60$ A lagging 120° with a power factor of 0.9

3. A 230 V, 50 Hz operating coil takes 70 mA at a lagging power factor of 0.5. Using a phasor diagram determine the current taken from the supply if a 1 μF capacitor is connected across its terminals.

4. A balanced three phase load supplied at 400 V, 50 Hz consists of three similar single phase loads. Each load comprises an inductive coil of 20 Ω resistance and 0.15 H inductance and connected in series with a 120 μF capacitor. Calculate the line current, total power and power factor when the loads are connected
   (a) in star
   (b) in delta.

4

Power Factor

Complete the following to remind yourself of some important facts on this subject that you should remember from the previous chapter.

1. Solve for R_2 and V_2,

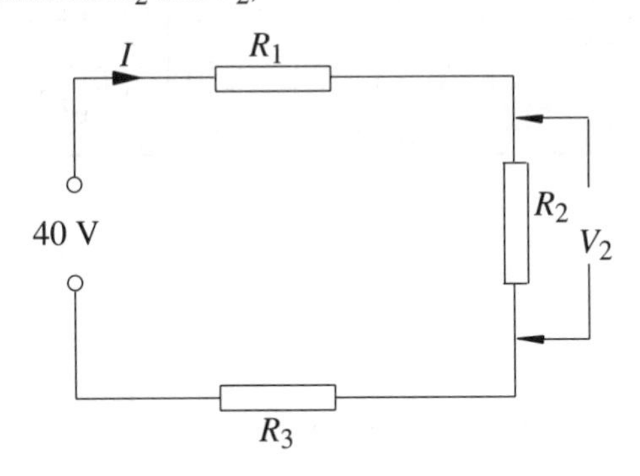

if $R_1 = 16\ \Omega$, $R_3 = 24\ \Omega$, and $I = 250$ mA.

2. Draw to scale, the impedance and power triangles for the following circuit:

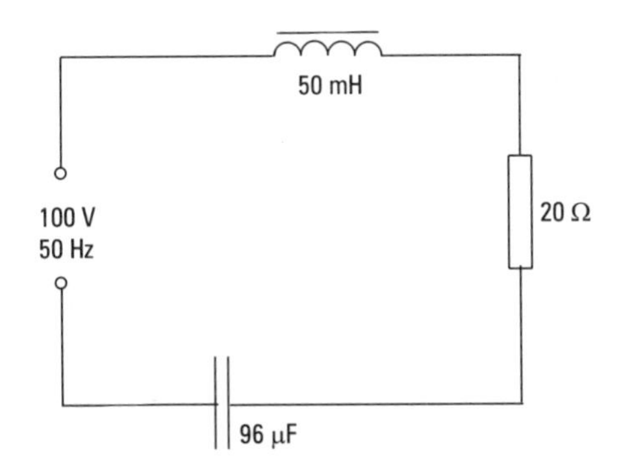

3. What is the current drawn by a 400 V, 40 kW 3 phase load having a power factor of 0.8?

On completion of this chapter you should be able to:

- plot the relationships between "in-phase" and "out of phase" waveforms
- calculate true power and VoltAmps in circuits supplied with a.c.
- determine the power factor for circuits containing resistance and reactance
- describe methods of correcting the power factor on three-phase loads
- describe the equipment used for power factor correction

The purpose of this chapter is to take a closer look at power factor. We will try to define what it is, what causes it and what can be done about it.

So here goes!

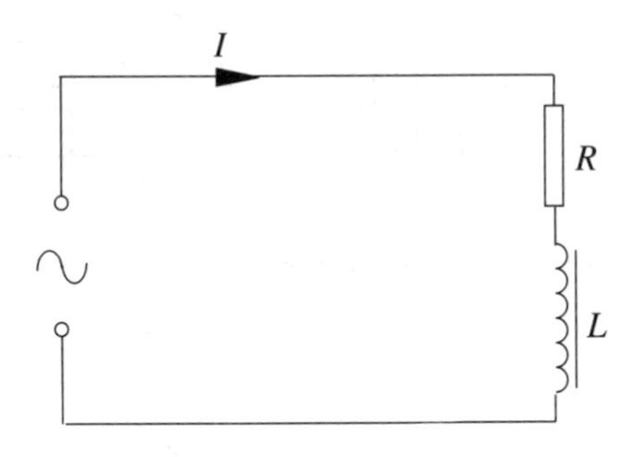

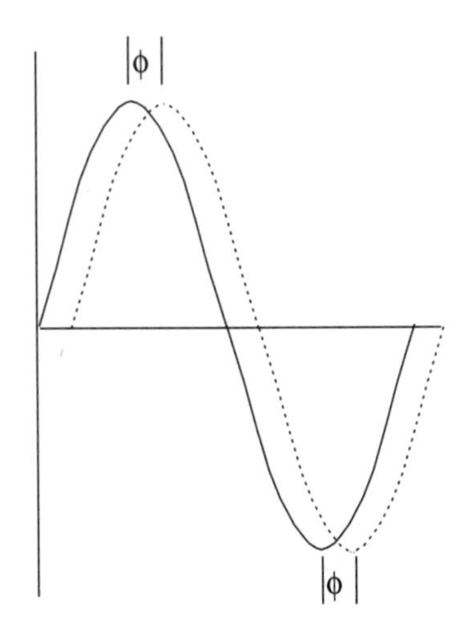

Figure 4.1

In its simplest form, an alternating voltage is sinusoidal. That is to say, it rises and falls in a regular manner so that the waveform follows a precise pattern in which the amplitude (the height) is proportional to the SINE of an angle as it rotates through one complete revolution.

You can plot this on a piece of squared paper with the horizontal axis marked out in degrees from 0 to 360.

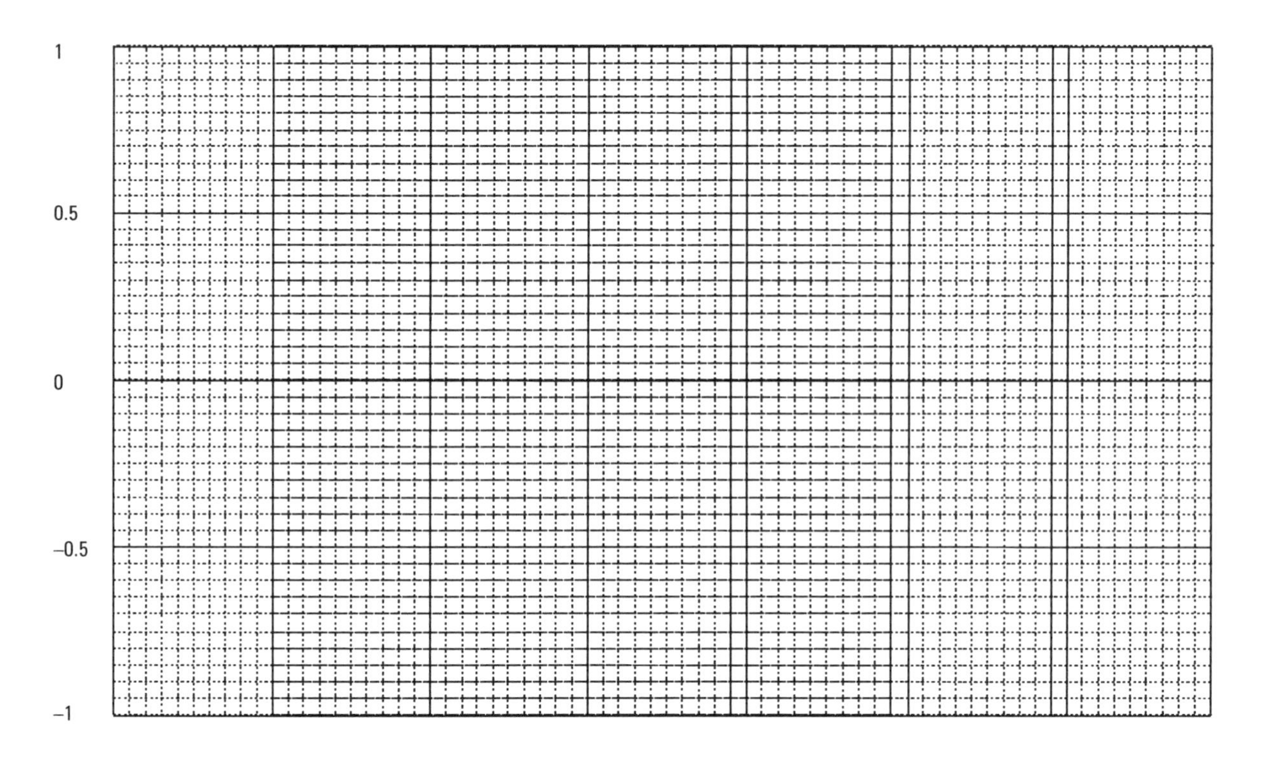

Using your calculator, write down the sine of the angle in steps of 15°.

Enter the angle on your calculator, then press the "sin" key and read off the value given.

Make sure that your calculator is set to degrees (or DEG) because if it is on Radians or GRAD it will NOT give the same results.

0	180 . . . .
15	195 . . . .
30	210 . . . .
45	225 . . . .
60	240 . . . .
75	255 . . . .
90	270 . . . .
105	285 . . . .
120	300 . . . .
135	315 . . . .
150	330 . . . .
165	345 . . . .
	360 . . . .

You will notice that the results follow a regular, repetitive pattern and you can predict what the values are going to be before you even press the sine key.

Join the points together with a smooth curve and you should have completed a pretty decent sine wave.

Power

If you look at two sine waves, one representing voltage and the other current, the power at any instant is the product of the current and the voltage at that instant.

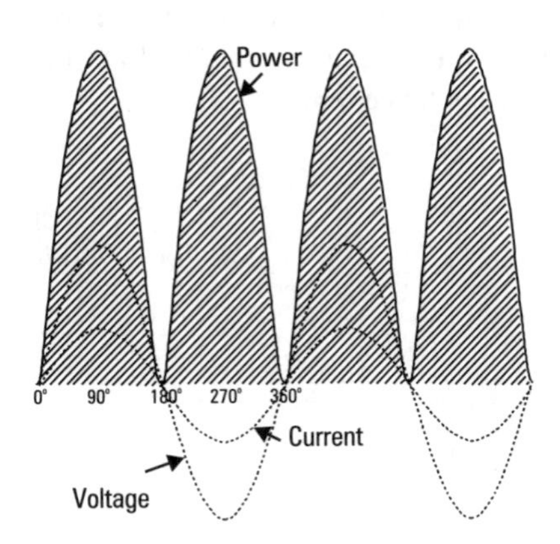

Figure 4.2 *Resultant power with voltage and current in phase*

You can see that all the power produced by the circuit is positive, i.e. " real" power. Even when current and voltage are negative, they are both negative at the same time and when multiplied together the result is a positive quantity.

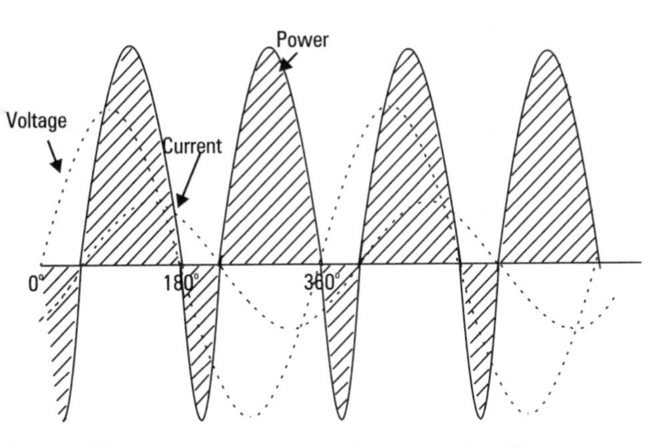

Figure 4.3 *Voltage leading the current by 60°*

Now look at the sine waves in Figures 4.2 and 4.3. They are identical in form and amplitude but the first one starts at 0° and the second passes through the axis at 60° and so they continue, rising and falling in exactly the same manner but out of step with each other.

When the current and voltage at any instant are multiplied together, the resultant power may be positive or negative depending on whether the current and voltage are the same side of the zero line.

To take matters a stage further:
When the current and the voltage are 90° out of phase there will be exactly the same amount of "negative" power as "positive" and the overall effect is that no power is produced in that circuit, regardless of the fact that current and voltage are present.

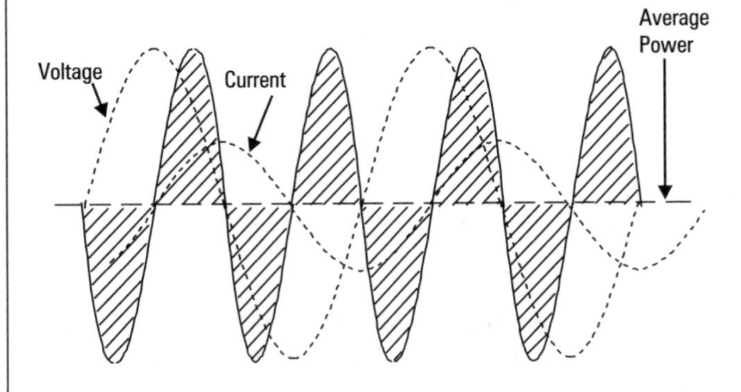

Figure 4.4 *Resultant negative and positive power when voltage leads the current by 90°*

This is just a graphical method of showing what you already know; namely, that power in an a.c. circuit can be determined by

$$P = U \times I \times \cos\phi$$

(where ϕ is the angle of phase difference between current and voltage.)

WHAT CAUSES IT ?

In a purely resistive circuit an alternating current flows through the resistor and the power at any instant can be found by $P = I^2 R$ Watts, so the mean value of power using r.m.s. values can be written

$$I^2 R = I \times I R$$

But IR = Volts and I = Amps

Therefore the power in a purely resistive circuit can be found by multiplying together the voltmeter and ammeter readings.

i.e. Watts = volt amps

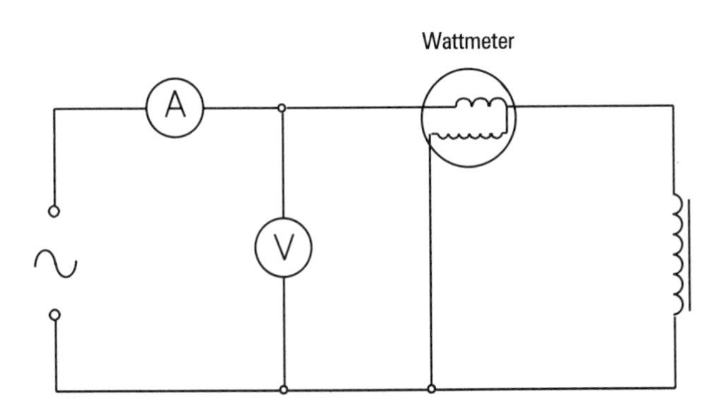

Figure 4.5 *Measurement of voltage, current and watts*

If it were possible to create a circuit which was purely inductive, the current flowing would depend on the voltage and the inductive reactance only. A current will flow round the circuit as before, but in this case it will not be in step with the voltage. The reason for this is the current in the inductor

$$I = \frac{U}{X_L}$$

You will recall from previous examples that

$$X_L = 2\pi fL$$

The current in the circuit would be sinusoidal, i.e. it would follow a sine wave form and its direction of flow would alternate.

The basic principle of self inductance tells us, that in an inductive circuit there will be an e.m.f. induced in the inductor which will oppose the change in current. The magnitude of this induced e.m.f. will be proportional to the rate of change of current.

If we look at the sine waves of current and induced e.m.f. you will see that the maximum negative e.m.f. occurs when the current is changing at its greatest rate, i.e. when the current is changing from positive to negative before increasing towards its maximum positive value.

At the top of the curve, the rate of change slows down and momentarily stops changing before starting to decrease. This is accompanied by an induced e.m.f. which is proportional to the rate of change and opposing it as implied by Lenz's Law.

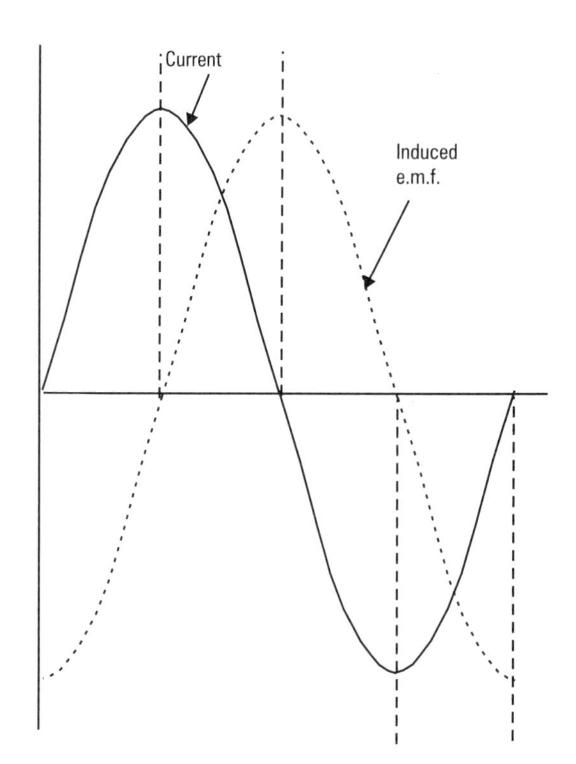

Figure 4.6 *Induced e.m.f. lagging behind the current*

Throughout the negative half cycle of the current, the induced e.m.f. behaves in a predictable manner and opposes the decrease in current. As you can see from the current and induced e.m.f waveforms, the current leads the induced e.m.f by a quarter of a cycle (Figure 4.6).

As it has been assumed that there is no resistance in this circuit, the induced e.m.f. is purely REACTIVE i.e. equal and opposite to the applied e.m.f.

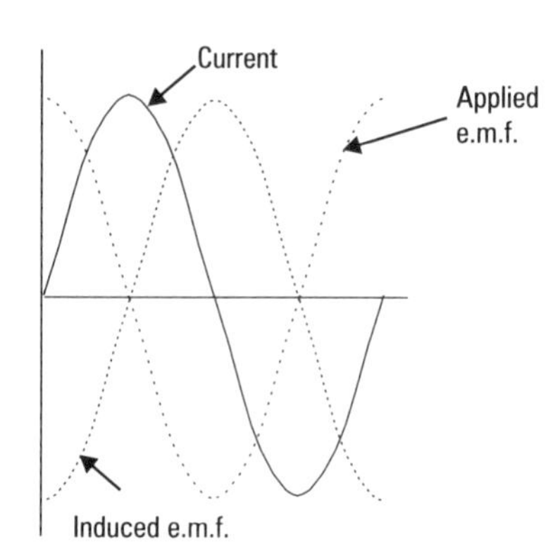

Figure 4.7 *The induced e.m.f. lags behind the current which is lagging behind the applied voltage*

In Figure 4.7 you will see that current lags the applied e.m.f. by 90° in a purely inductive circuit.

In practical terms, it is likely that you will be working with inductors which have a significant amount of resistance so therefore a phase angle of 90° and a corresponding power factor of 0 is a rather theoretical situation.

When a circuit contains resistance and inductance the effect will be reduced so that the phase angle lies somewhere between 0° and 90° and the power factor between 1 and 0 depending on the cosine of the phase angle.

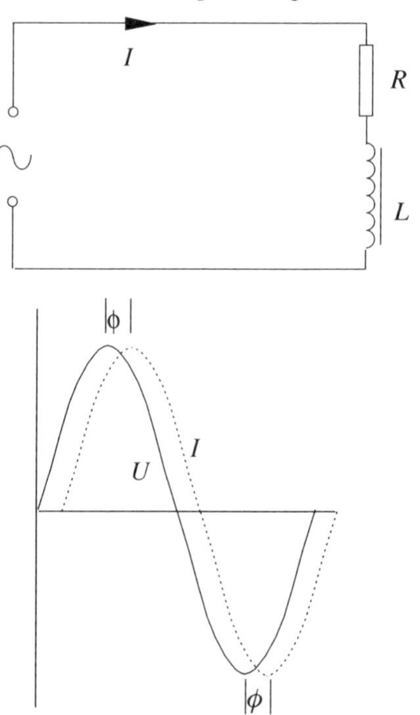

Figure 4.8 *Current lags behind the applied voltage due to the inductive reactance in the circuit*

Capacitance in a.c. circuits

As the potential difference across the plates of a capacitor increases so current will flow into the capacitor to increase its charge according to the increased voltage. If we assume that there is to be no resistance in this circuit then this response will be immediate.

From this it can be said that the current at any instant will be proportional to the rate of change of voltage.

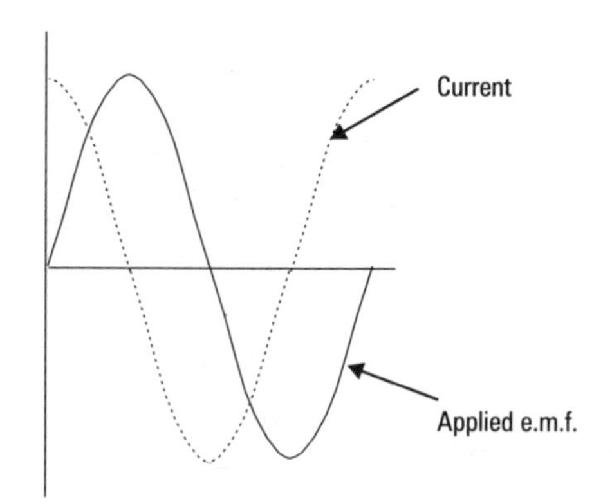

Figure 4.9 *Current leads applied voltage*

Figure 4.9 shows that when the applied voltage is changing, the capacitor current is at its greatest and when the current ceases to change (at the top and bottom turning points) the capacitor current is zero.

The effect of this is to create a situation where the alternating current in a purely capacitive circuit leads the applied e.m.f. by 90° and the power factor is zero.

This is described as a " leading" power factor and its effect is the opposite of that found in the purely inductive circuit.

In a situation which is not purely capacitive but has a certain amount of resistance, the phase angle (ϕ) will be less than 90° and the power factor somewhere between 1 and 0.

In terms of the power dissipated in the circuit the result will be the same namely:

Power in Watts

$$P = U \times I \times \cos\phi$$

The current drawn by a capacitor in an a.c. circuit is found by

$$I = \frac{U}{X_C} \qquad \text{where} \qquad X_C = \frac{1}{2\pi fC}$$

Power factor improvement

An installation having a low power factor is a liability to the consumer and the supplier.

The consumer will be paying more for electricity because of penalty charges levied by the supplier on customers who have a low power factor and maximum demand charges based on the kVA rating of the load would add to his energy costs. In addition to this the installation cables may have to be larger than strictly necessary because they are carrying more current due to the power factor.

The supplier finds low power factor a burden because the additional current loads up generating plant, cables, transformers and switchgear without supplying any additional power.

Low power factor means wasted resources and bad economics. Power factor improvement, properly applied, repays any capital investment many times over.

Example

A consumer has a maximum demand of 150 kW at 400 V, 3 phase with a power factor of 0.57. The maximum demand kVA would be

$$\text{kVA} = \frac{\text{kW}}{\cos\phi}$$

$$= \frac{150}{0.57}$$

$$= 263 \text{ kVA}$$

This means that the installation would require transformer capacity of at least 263 kVA plus all the associated switchgear of a similar rating.

In terms of current, the line current at 400 V would be

$$I_L = \frac{P}{\sqrt{3}\, U_L \cos\phi}$$

$$= \frac{150000}{\sqrt{3} \times 400 \times 0.57}$$

$$= 380 \text{ A}$$

This would require cables of a suitable rating, added to which would be the cost of the terminations and the distribution equipment needed to handle such a large current.

At unity power factor, the kVA would be equal to the kilowatts i.e. 150 kW = 150 kVA. This would mean a considerable reduction in capacity.

Also at unity power factor:

$$I_L = \frac{P}{\sqrt{3}\, U_L \cos\phi}$$

$$= \frac{150000}{\sqrt{3} \times 400 \times 1}$$

$$= 216.51 \text{ A}$$

which represents a quite significant reduction in current.

This reduction is achieved without any loss of power to the consumer. It can be achieved by installing the appropriate power factor improvement equipment at a modest capital outlay and this can be recovered from the reduction in operating costs.

HOW DOES IT WORK?

Look at the phasor addition of two currents.

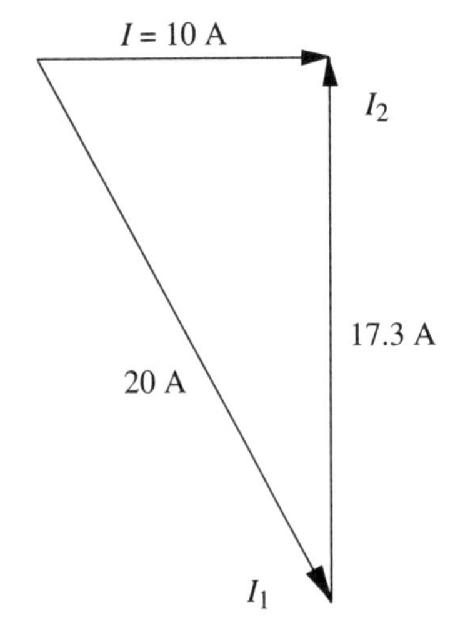

Figure 4.10

I_1 is 20 A lagging by 60°

I_2 is 17.3 A leading by 90°

The phasor sum is the resultant phasor, which you will see is 10 A.

In fact, the two currents added together, produce a total which is smaller than either of them. This is not unusual and similar situations are to be found in many discharge lighting circuits.

Example

A large coil has a resistance of 10 Ωand an inductive reactance of 20 Ω. If this coil is to be connected to a 100 V 50 Hz. a.c. supply determine the

(a) impedance
(b) current drawn from the supply
(c) power factor of the coil

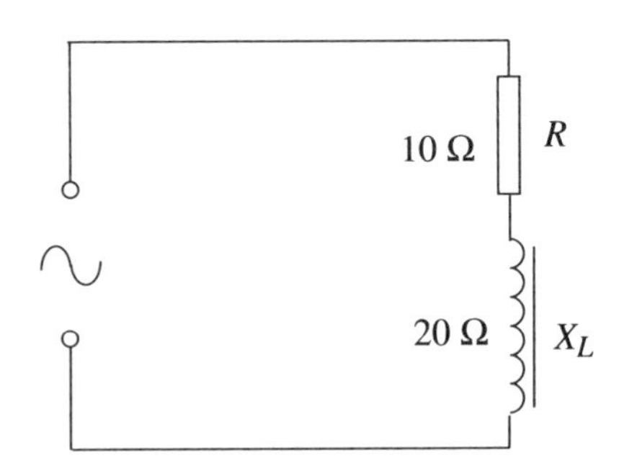

Figure 4.11

(a) $Z = \sqrt{R^2 + X_L^2}$
$= 22.36\ \Omega$

(b) $I = \frac{U}{Z}$
$= 4.47 \text{ A}$

(c) $\cos\phi = \frac{R}{Z}$
$= 0.447$

If we now look at the circuit in this example, Figure 4.12, which has had a 127.3 μF capacitor added to it, we will see what effect the addition of this capacitor has had on the power factor and on the current taken from the supply.

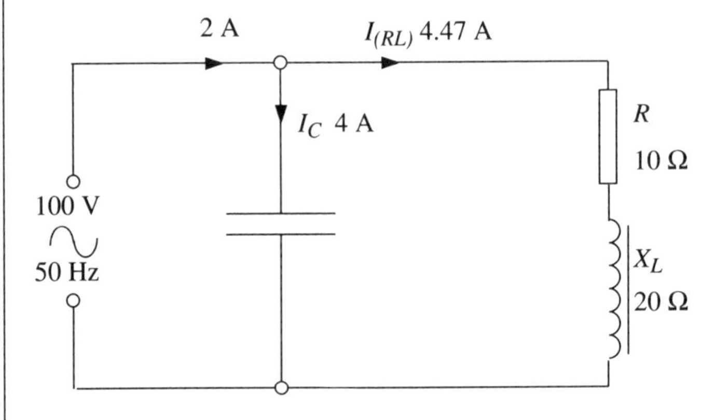

Figure 4.12

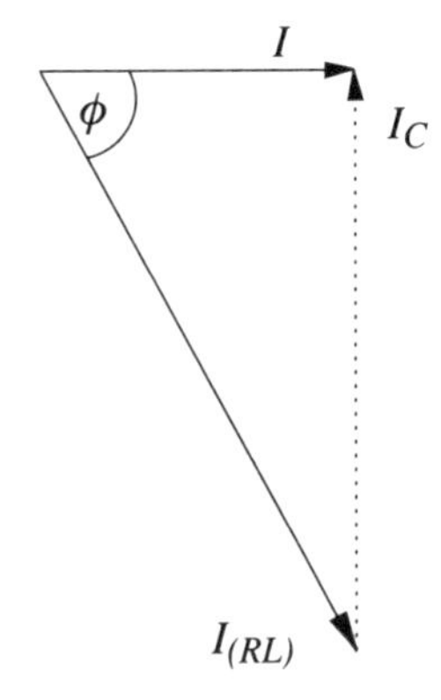

Figure 4.13

$$I_C = \frac{U}{X_C}$$

$$X_C = \frac{1}{2\pi fC}$$

$$= 25\ \Omega$$

$$I_C = \frac{100}{25}$$

$$= 4\ \text{A}$$

From the diagram you see how a current of 4.47 A, lagging by an angle of 63.4° is corrected to a current of 2 A at unity by the connection of a parallel capacitor of the appropriate value.

Try this

An 80 V 50 Hz a.c. circuit has a resistance of 16 Ω and an inductive reactance of 12 Ω. A capacitor of 99.5 μF is connected in parallel with the circuit.

Determine:

(a) the impedance of the coil
(b) the current taken by the coil
(c) the power factor of the coil
(d) the reactance of the capacitor
(e) the current taken by the capacitor
(f) draw a scaled phasor diagram of the currents to show how the capacitor has corrected the power factor.

It is also possible to look at power factor correction without considering the currents.

We are already familiar with the technique of representing an a.c. load in terms of Power, VoltAmps and VoltAmps Reactive.

Example

A 10 kVA load is made up of 6 kW and 8 kVAr and is lagging by an angle of 53.13°.

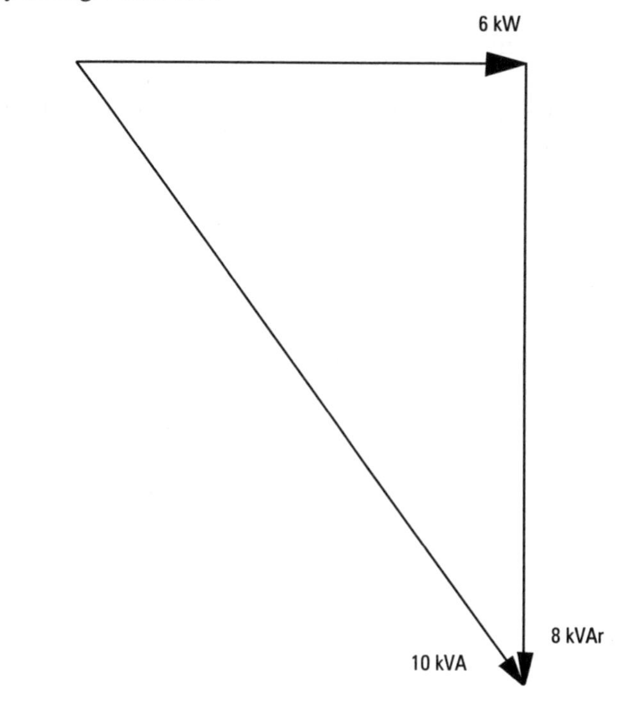

Figure 4.14 Phasor diagram

In the simplest possible terms, all that is required to correct the power factor to unity is a capacitor of 8 kVAr.

Three-phase loads

Power factor correction of three-phase balanced loads need be no more difficult than the last example.

Example

A 24 kW, three phase balanced load has a power factor of 0.8. How many kVAr of capacitance is required to correct the power factor to unity?

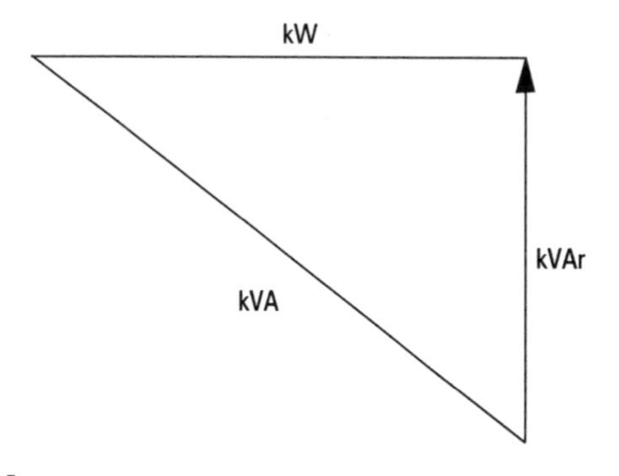

Figure 4.15

$$kW = 24$$

$$kVA = \frac{24}{\cos\theta}$$

kVAr (by graphical solution) = 18 kVAr

This solution tells us that a balanced three-phase arrangement of capacitors with a total reactive load of 18 kVAr would be sufficient to correct the power factor to unity.

Alternatively:

The problem could be solved by using trigonometry only, as follows:

$$\text{Power} = UI\cos\phi$$
$$\phi = \text{inv}\cos 0.8$$
$$= 36.87°$$

$$\sin\phi = 0.6$$
$$\text{Power} = 24\text{ kW}$$

$$\text{kVA} = \frac{24}{\cos\phi}$$

$$= 30\text{ kVA}$$

$$\text{kVAr} = UI\sin\phi$$
$$= 30 \times 0.6$$
$$= 18\text{ kVAr}$$

Try this

How many kVAr of capacitance is required in order to correct to unity, a balanced three phase load of 11.3 kW at a power factor of 0.68?

Star and delta connected three phase capacitors

Star connection

Take for example, a three phase star connected capacitor with a total kVAr of 18.

This represents a balanced three phase capacitance of 6kVAr per phase.

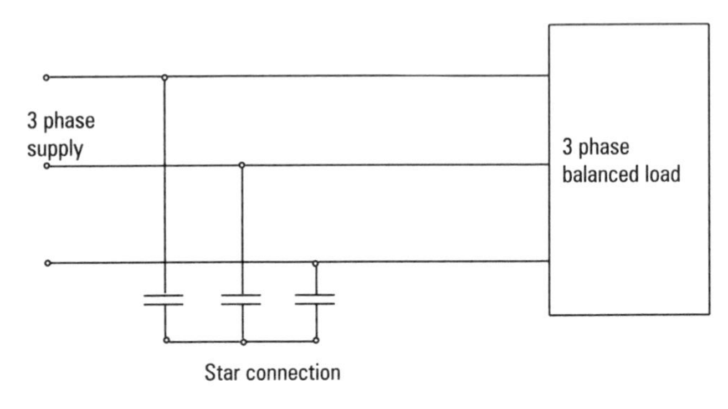

Figure 4.16 Capacitors connected in star

Assuming that this is to be connected to a 50 Hz 400/230 V system, in STAR connection, the voltage per capacitor is 230 V.

For a balanced capacitive load of this nature, the line current would be:

$$I_C = \frac{VA_r}{\sqrt{3} \times U_L}$$

$$= \frac{18000}{692.8}$$

$$= 26\text{ A (very nearly)}$$

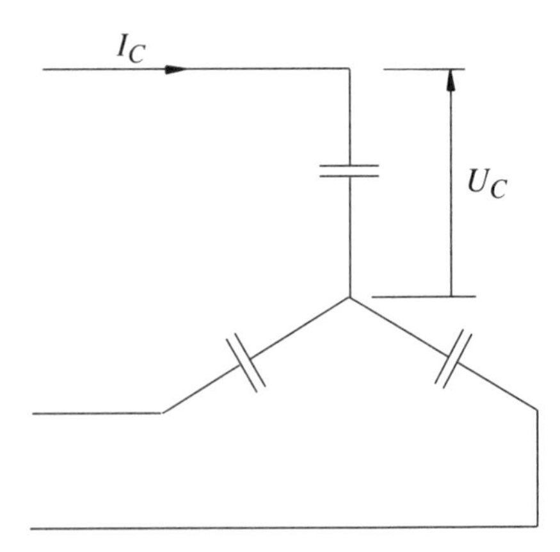

Figure 4.17

The voltage per capacitor is 230V, therefore the reactance is

$$\frac{230}{26} = 8.85\ \Omega$$

Since $$X_C = \frac{1}{2\pi fC}$$

then $$C = \frac{1}{2\pi fX_C}$$

giving a capacitance per phase of 359.67 μF.

Delta connection

If we take the previous example but use a DELTA connected capacitor arrangement.

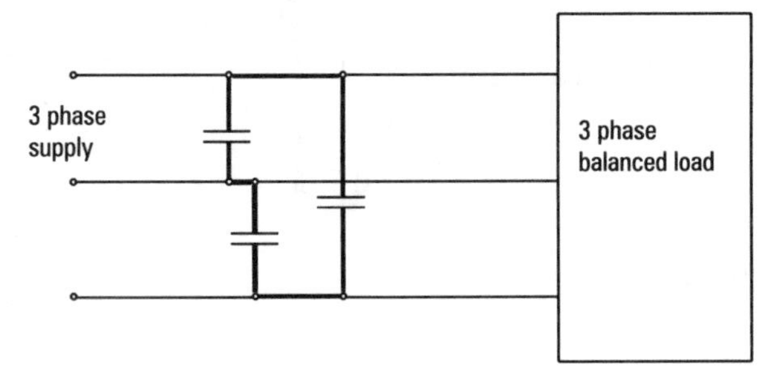

Figure 4.18 Delta connected capacitors

The voltage per capacitor is now 400 V.

The total kVAr is still 18, therefore the line current is still 26 A.

In a delta connected load the phase current

$$I_P = \frac{I_L}{\sqrt{3}}$$

this gives a phase current of

$$= \frac{26}{\sqrt{3}}$$

$$= 15\ \text{A}$$

making the reactance per capacitor

$$X_C = \frac{400}{15}$$

$$= 26.66\ \Omega$$

Which gives a capacitance per phase of

$$= \frac{1}{2\pi fX_C} = 119.36\ \mu\text{F}$$

Example

A three phase, 400 V 50 Hz 10.3 kW balanced load has a power factor of 0.72. What is the capacitance per phase required to correct the power factor to unity?

(a) in STAR

(b) in DELTA

Solution

$$\text{Total kVA} = \frac{\text{Total kW}}{\text{power factor}}$$

$$= \frac{10.3}{0.72}$$

$$= 14.3\ \text{kVA}$$

$$\phi = \text{inv cos } 0.72 \qquad (\cos^{-1} 0.72)$$

$$= 43.9°$$

$$\sin\phi = 0.694$$

Total reactive component

$$\text{kVAr} = \text{kVA} \times \sin\phi$$

$$= 9.92\ \text{kVAr}$$

Capacitive current per phase

$$I_C = \frac{\text{VA}_r}{\sqrt{3} \times U_L}$$

$$= \frac{9920}{692.8}$$

$$= 14.32\ \text{A}$$

(a) In STAR connection

Current per phase

$$I_P = I_L = 14.32\ \text{A}$$

Voltage per phase

$$U_P = \frac{U_L}{\sqrt{3}}$$

$$= 230\ \text{V}$$

Capacitive reactance

$$X_C = \frac{U_C}{I_C}$$

$$= \frac{230}{14.32} = 16.06\ \Omega$$

$$\text{Capacitance} = \frac{1}{2\pi fX_C}$$

$$= 198.18\ \mu\text{F}$$

(b) DELTA connection

$$I_P = \frac{I_L}{\sqrt{3}}$$

Capacitor current per phase

$$I_C = \frac{14.32}{\sqrt{3}}$$

$$= 8.27 \text{ A}$$

Capacitor voltage per phase

$$= U_L$$

$$= 400 \text{ V}$$

Capacitive reactance

$$X_C = \frac{400}{8.27}$$

$$= 48.37\ \Omega$$

Capacitance per phase

$$C = \frac{1}{2\pi f X_C}$$

$$= 65.8\ \mu\text{F}$$

Try this

Determine the capacitance per phase required to correct the power factor to unity if a 400 V, 12 kW balanced three phase load has a lagging power factor of 0.556 at a frequency of 50 Hz:

(a) in STAR
(b) in DELTA

Correction to power factors other than unity

It is not always necessary to correct a lagging power factor to unity, and the economics of power factor improvement may not require a power factor of more than 0.9 or 0.95 at the most.

We will now take a look at power factor "improvement" as opposed to power factor "correction".

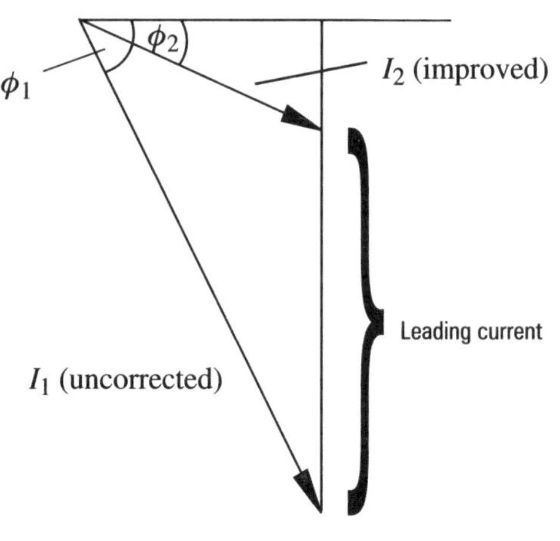

Figure 4.19

The graphical method

Take the example of a single phase load which has a current of 60 A at a power factor of 0.5 and how this current can be reduced by improving the power factor to 0.9.

$$\cos\phi_1 = 0.5$$
$$\phi_1 = 60^\circ \text{ lagging}$$
$$\cos\phi_2 = 0.9$$
$$\phi_2 = 25.8^\circ$$

1. By constructing a scaled phasor diagram **A B C** we can show the current I_1 and its phase angle at a power factor of 0.5, (**A C**) (Figure 4.20).
2. By constructing a second phasor diagram **A B D** inside the first we can show the the current I_2 and its phase angle at 0.9 (**A D**).

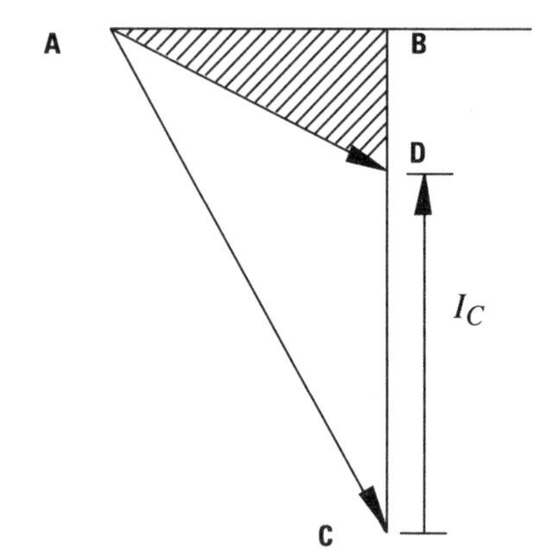

Figure 4.20

The significant differences are:

(a) The reduction in load current from 60 A (**A C**) to 33.3 A (**A D**)

(b) The shortening of the vertical axis by 37.45 A

The length **CD** is equivalent to the leading current (I_C) drawn by a capacitor which compensates for a large part of the lagging component **BC**.

To take the matter a stage further; if we assume the supply to be 230 V 50 Hz, this would be the current drawn by a capacitor with a reactance of

$$X_C = \frac{U}{I_C}$$

$$= \frac{230}{37.45}$$

$$= 6.14\ \Omega$$

This in turn relates to a capacitor with a capacitance of

$$C = \frac{1}{2\pi f X_C}$$

$$= 518.29\ \mu F$$

Example

A 230 V 50 Hz single phase load takes a current of 12 A at a power factor of 0.6. What value of capacitance is required to improve the power factor to 0.95?

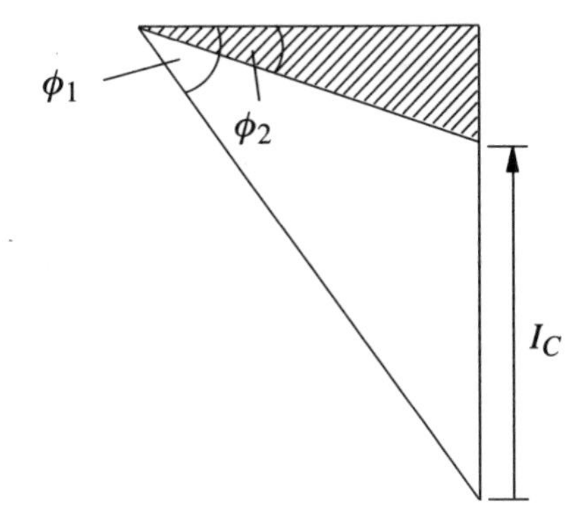

Figure 4.21

$$\cos\phi_1 = 0.6$$
$$\phi_1 = 53.1°$$
$$\cos\phi_2 = 0.95$$
$$\phi_2 = 18.9°$$

$$I_C = 7.24\ \text{A (from Figure 4.21)}$$
$$X_C = \frac{U_C}{I_C}$$
$$= 31.76\ \Omega$$

$$C = \frac{1}{2\pi f X_C} = 100.1\ \mu F$$

Try this

Determine, by the graphical method, the value of capacitance required to improve the power factor of a 230 V single phase 50 Hz load, taking a current of 26 A at a lagging power factor of 0.44 to a power factor of 0.8.

Power factor improvement by determining the required kVAr

It is also possible to use the same method by considering the reactive component of the relative power triangles.

Example

A 20 kVA load has a power factor of 0.65 and this is to be improved to 0.85 by means of a capacitor.

$$\cos\phi_1 = 0.65$$
$$\phi_1 = 49.45°$$
$$\cos\phi_2 = 0.85$$
$$\phi_2 = 31.78°$$

Construct one triangle inside the other as before.

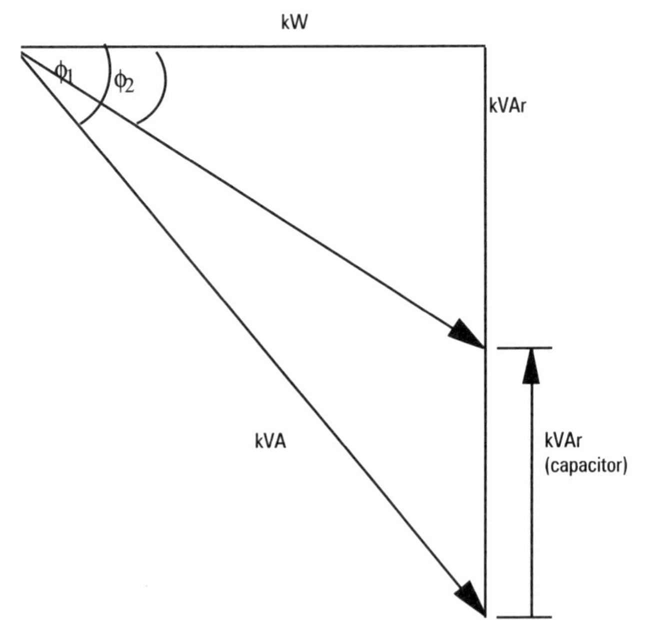

Figure 4.22

Determine the difference between the two kVAr values by measurement and this represents the required value of kVAr of the capacitor.

If the value of capacitance is required this can be derived from the kVAr as before.

i.e.

$$I_C = \frac{\text{VAr}}{U_C}$$

$$X_C = \frac{U_C}{I_C}$$

$$C = \frac{1}{2\pi f X_C}$$

Try this

A load of 75 kVA has a power factor of 0.55. Determine the kVAr of the capacitor required to improve the power factor to 0.9.

Three phase

Three phase loads are no more of a problem. Simply determine the overall kVAr required by the same method and since we are only considering balanced loads, divide the total by three to find the kVAr per phase.

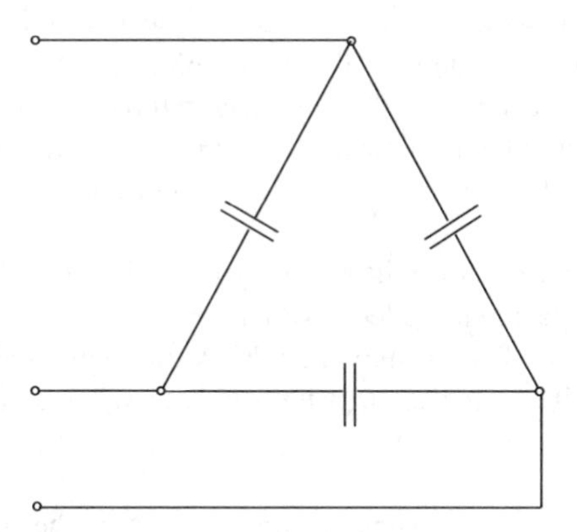

Figure 4.23 *Delta connected capacitors for power factor improvement*

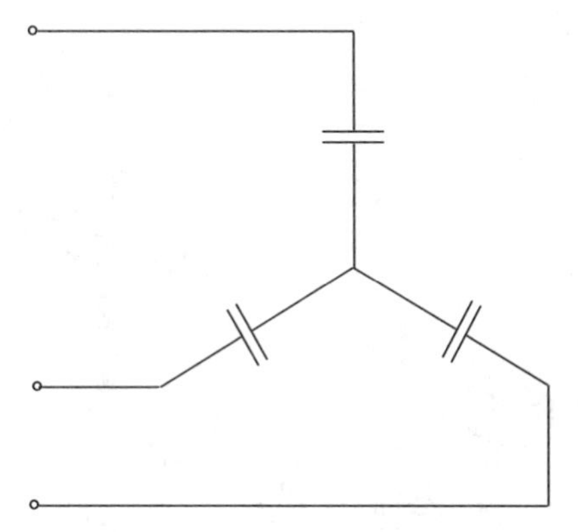

Figure 4.24 *Power factor improvement using star connected capacitors*

Example

It is required to improve the power factor of a 30 kVA three phase load from 0.5 to 0.85 by the application of a power factor improving capacitor. How many kVAr will be required per phase?

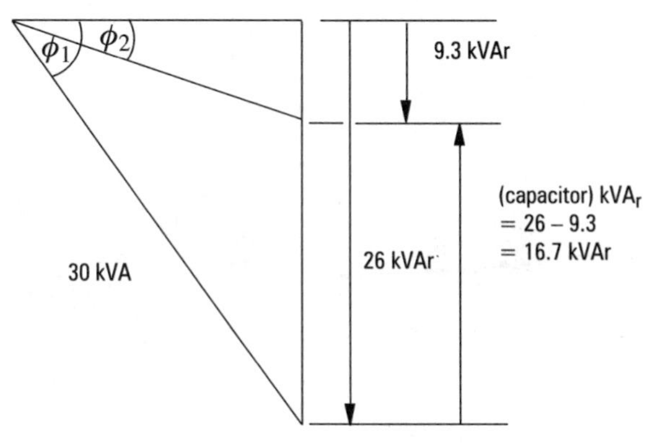

Figure 4.25

$$\cos\phi_1 = 0.5$$
$$\phi_1 = 60^\circ \qquad \sin\phi_1 = 0.866$$
$$\cos\phi_2 = 0.85$$
$$\phi_2 = 31.78^\circ \qquad \sin\phi_2 = 0.527$$

$$\sin\phi_1 - \sin\phi_2 = 0.339$$

In a 30 kVAr load, this represents a capacitive reactance of 16.7 kVAr (from Figure 4.25)

For a balanced three phase load, this would be $\frac{16.7}{3}$ i.e. 5.57 kVAr of capacitance per phase

Remember

You will, by now, have seen sufficient examples of power factor correction to have got the general idea.

A graphical solution will give an acceptable answer in most cases and is probably the easiest method provided that you use basic measuring and drawing instruments and apply a bit of care in the construction of the figures. For the more mathematically minded, a trigonometrical solution will prove to be quicker and more accurate.

Solution by either method is usually acceptable to examiners in the subject so you would be well advised to use the method which suits you best.

Power factor correction equipment – other than capacitors

Capacitors are more widely used than any other form of power factor improvement. This does not mean that all other forms are excluded. Any item of a.c. plant having a leading power factor will improve the overall power factor of a lagging installation.

A classic example of this type of plant is the synchronous a.c. motor, an item of plant which can be used for a functional purpose as a motor as well as power factor improvement.

The synchronous motor is quite different from conventional induction motors in that it runs at synchronous speed and has rotor windings which are supplied with a d.c. excitation current.

The power factor of this type of machine is variable over a wide range from lagging through unity and into leading by adjusting the excitation current. This means that the amount of power factor correction can be adjusted to suit load conditions.

The motor will, in the meantime, be used for some other useful purpose such as driving a fan or circulating pump or any other constant load which is left running continuously.

A fuller account of the characteristics of this type of machine will be found in the studybook on "ELECTRICAL MACHINES" which forms part of this series.

The connection of power factor correction equipment

For the best possible effect, the equipment should be connected to the actual load. This means that the cables supplying the load will have the benefit of power factor correction and the final circuit current will be less as a result.

This is the technique which is commonly used in discharge lighting circuits.

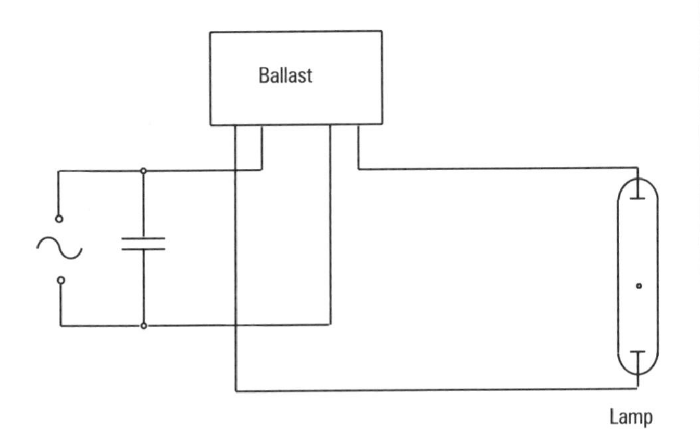

Figure 4.26 *Power factor correction capacitor connected into a discharge lighting circuit*

For the overall power factor correction of a complete installation, power factor correction capacitors are connected to the busbars of the supply intake equipment. This does nothing for the internal wiring of the installation but does have the desired effect on the whole plant thus reducing the consumer's energy costs by avoiding the penalty charges levied by the supplier.

The economics of power factor correction

Modern power factor correction equipment is supplied in self-contained, wall or floor mounting units with all the necessary control and protection devices incorporated. These units may be manually or automatically operated and require very little maintenance once installed. The capacitors of a modern unit are normally of the dry film type which is non-toxic and they do not use the type of PCB dielectric incorporated in older units which can cause serious health and environmental hazards and makes the disposal of redundant plant a very difficult and costly process.

Good quality equipment of this type is not cheap and it must be correctly installed if it is to function effectively but the cost advantage to the consumer is quite considerable.

A medium-sized industrial consumer who installs a power factor correction unit of 50 to 60 kVAr can expect to recover the capital cost of its installation within three to four years and for the rest of the lifetime of the unit is making real savings in the cost of his electrical energy consumption.

A low power factor will normally result in a penalty charge based on that alone, added to which, a maximum demand tariff which is based on the maximum kVA supplied within a given period will be unnecessarily high and this will result in higher charges.

Power factor improvement does not affect the amount of power delivered to the consumer but it does affect the overall efficiency and the economy of operating an electrical installation.

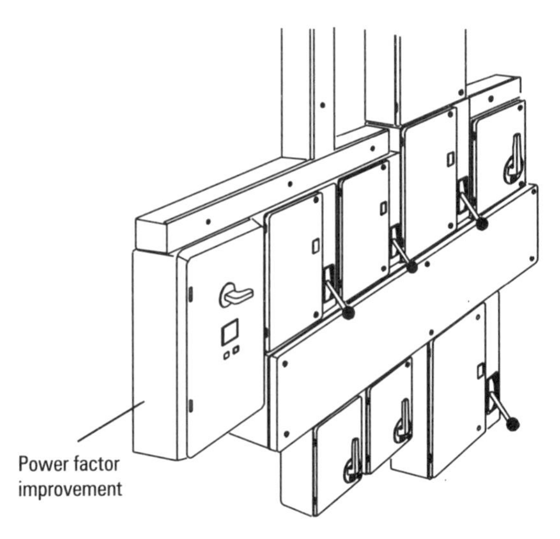

Figure 4.27

Exercises

1. (a) State TWO methods which may be used to improve the power factor of an industrial installation.
   (b) A factory with an overall load of 400 kW and a power factor of 0.7 lagging is required to improve its power factor to 0.85. The supply is 400 V, three phase 50 Hz.

Determine
   i) the required kVAr per phase
   ii) the required capacitance per phase for DELTA connected capacitors.

2. (a) Describe THREE methods by which the power factor of a large industrial installation may be improved.
   (b) A factory with a total load of 500 kVA is required to improve its power factor from 0.7 to 0.95. If the supply is 6.6 kV three-phase 50 Hz, determine the value of each of the three DELTA connected capacitors required for this purpose.

3. (a) With the aid of a fully labelled diagram, explain how the power factor of a large three-phase motor may be improved by the connection of capacitors.
   (b) Instruments connected to the circuit show that the total power supplied to the load is 20 kW whilst the line current is 52 A at a line voltage of 400 V. Find the capacitance per phase required to improve the power factor to 0.9.

4. (a) State THREE advantages of power factor correction.
   (b) Show, by means of sketches, how three capacitors can be connected in order to improve the capacitance of a three-phase industrial load
   i) in star
   ii) in delta
   (c) A factory has a total load of 150 kVA at a power factor of 0.65 lagging. Determine the kVAr per phase of capacitance required to improve this power factor to 0.95.

5

Transformers

Complete the following to remind yourself of some important facts on this subject that you should remember from the previous chapter.

1. Low power factor can be a burden because the additional current loads up generating plant, cables, transformers and switchgear without ____________ ____________ ____________ ____________.
   Low power factor means wasted ____________ and bad ____________________.

2. A 6 kW, 3 phase balanced load has a power factor of 0.6. How many kVAr of capacitance is required to correct the power factor to unity?

3. A load of 25 kVA has a power factor of 0.65. What is the kVAr of the capacitor required to improve the power factor to 0.85?

On completion of this chapter you should be able to:

- describe the relationship between voltages, currents and number of turns in double wound transformers
- calculate the "input" and "output" currents and voltages of different transformers
- calculate the efficiency of transformers
- describe the different types of losses
- determine the regulation and impedance
- identify different methods of winding transformer coils
- identify transformer connections from their terminal markings

The transformer is an extremely useful piece of electrical equipment. Its main use is to take an a.c. supply at one voltage and produce from this another a.c. supply. The voltage of the second supply may be quite different from that of the first and the process can isolate one from the other, a feature which makes the transformer ideal for the provision of safety services.

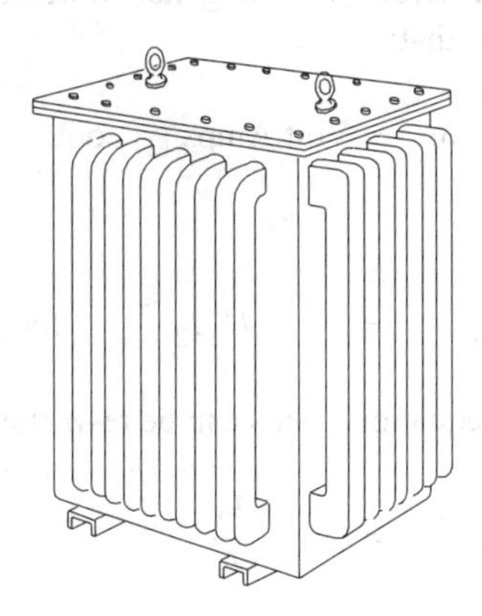

Figure 5.1

The transformer is a static electromagnetic device which operates on the principle of MUTUAL INDUCTANCE. Having studied the chapter on electromagnetism, you will be aware that a coil which is situated in its own magnetic field is capable of producing an induced e.m.f. due to its inductive properties.

If two coils share the same magnetic field, an e.m.f. will be induced in both coils. When a change in current in the first coil induces an e.m.f. in the second they are said to be mutually inductive.

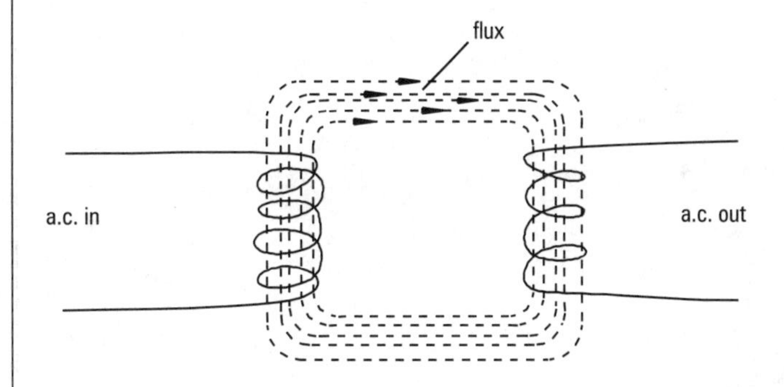

Figure 5.2

For all practical purposes, a transformer will consist of two coils of insulated wire wound around the same magnetic core.

One coil, which will be called the "primary" will be connected to an a.c. supply. The alternating current in the primary coil will set up an alternating magnetic flux in the core and this will link with the turns of the "secondary" winding. The alternating voltage thus produced will depend on the number of turns of wire in the secondary, just as the strength of the field will depend on the number of turns in the primary.

Assuming for the moment that there are to be no losses in the process then the Ampere-turns in the primary winding will be equal to the Ampere-turns in the secondary.

(if we use the symbol N for the number of turns)

$$N_1 I_1 = N_2 I_2 \quad \text{(equation 1)}$$

On the same theme, and assuming that the device has no losses then we can say that:

Volt Amps in = Volt Amps out

In other words

$$U_1 I_1 = U_2 I_2 \quad \text{(equation 2)}$$

Going back to equation 1, this can be re-written as

$$\frac{N_1}{N_2} = \frac{I_2}{I_1}$$

and similarly, Equation 2 can be written:

$$\frac{U_1}{U_2} = \frac{I_2}{I_1}$$

Which gives us:

$$\frac{U_1}{U_2} = \frac{N_1}{N_2} = \frac{I_2}{I_1}$$

This is often referred to as
THE BASIC TRANSFORMER EQUATION

Try this

Transpose each of the following equations for the one shown in bold type.

(a) $\frac{\mathbf{N_2}}{N_2} = \frac{I_2}{I_1}$

(b) $\frac{U_1}{U_2} = \frac{\mathbf{I_2}}{I_1}$

(c) $\frac{N_1}{\mathbf{N_2}} = \frac{U_1}{U_2}$

(d) $\frac{I_2}{\mathbf{I_1}} = \frac{N_1}{N_2}$

Step down transformers

Transformers are widely used for electrical and electronic applications because they can change voltages from one level to another with relative ease.

If a transformer has 240 turns in the primary winding and 24 turns in the secondary it is said to have a TRANSFORMER RATIO of

10 : 1

In other words the primary voltage is reduced by a factor of 10.

This relationship can be expressed by the basic equation

$$\frac{N_1}{N_2} = \frac{U_1}{U_2}$$

Thus a transformer with a 10 to 1 ratio will have a secondary voltage which is relative to the primary namely:

$$U_2 = U_1 \times \frac{N_2}{N_1}$$

Taking the previous example:

If a transformer having 240 turns in the primary and 24 turns in the secondary is connected to a 125 V supply, the secondary voltage will be:

$$U_2 = U_1 \times \frac{N_2}{N_1}$$

$$= 125 \times \frac{24}{240}$$

$$= 12.5 \text{ V}$$

Step up transformers

A transformer can just as easily be used for the purpose of raising the voltage and where higher voltages are required for a particular application the transformer provides a quick and easy solution.

For example if an item of equipment which is mainly operated at 24V a.c. but for one particular process requires 600V then the solution could be the inclusion of a transformer having a step up ratio of 25 to 1.

Given that the primary winding is to contain 50 turns then the secondary would need to be 25 times greater.

$$U_2 = U_1 \times \frac{N_2}{N_1}$$

$$= 24 \times \frac{1250}{50}$$

$$= 600 \text{ V}$$

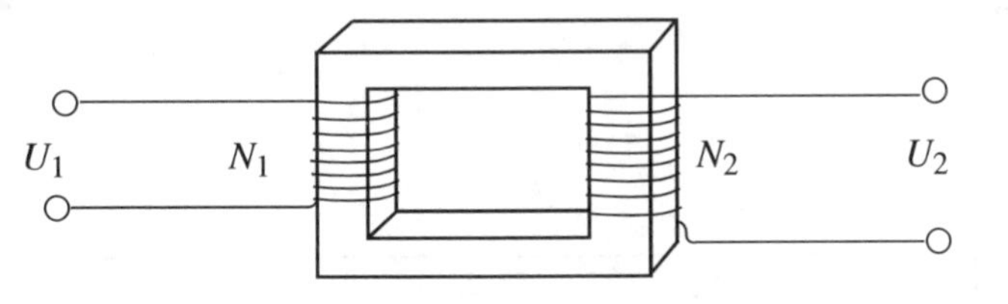

Figure 5.3

Isolating transformers

It is not always the case that transformers are used to change the voltage and for reasons of safety it may be required to provide a mains voltage supply which is not derived directly from the mains supply.

This is the principle adopted in the BS3535 bathroom shaver socket in which the 230 V shaver supply is provided by a socket outlet which has no reference connection to earth potential. It is therefore incapable of delivering an earth leakage current.

The same technique is used in the service and repair industries where technicians are frequently required to work on live electrical equipment.

If the equipment under repair is supplied through an isolating transformer then the technician, although still exposed to mains voltage, is not exposed to the dangers arising from simultaneous contact with exposed or extraneous conductive parts such as earthed apparatus or pipework.

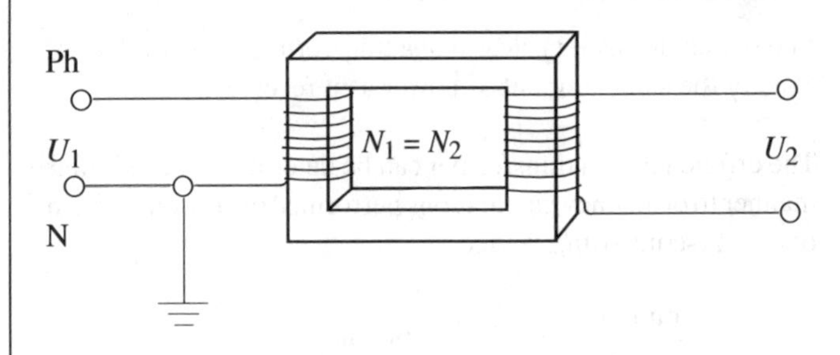

Figure 5.4 Isolating transformer with no connection to earth on the secondary winding.

Voltage/current relationships

Because the transformer is a very efficient item of equipment then for most practical purposes its efficiency can be assumed to be very close to 100%.

That is to say, the power delivered to the primary winding is assumed to be equal to the power delivered by the secondary.

i.e.

Power in = Power out

For example,

A 5 kVA transformer is supplied with 20 A at 250 V. Using this very basic relationship we could deduce that such a device would be capable of delivering 10 A at 500 V if the ratio happened to be 1:2.

Alternatively, if the secondary voltage happened to be 100 V the the current would have to be 50 A in order to maintain the same primary current and thus the ratio would be a step down in the order of 2.5:1.

Remember

Although transformers have no moving parts they are not 100% efficient.

Transformer efficiency

In practical terms it is clear that a transformer will not deliver exactly the same amount of power as it receives.

The efficiency of a transformer can be determined in a similar manner to efficiency calculations performed on any other form of energy-converting device.

$$\frac{\text{output}}{\text{input}} = \text{efficiency}$$

This is of course "per unit" efficiency and the result will be a fraction.

It is common practice to express efficiency as a percentage. This is nothing more than the per unit efficiency with the decimal point shifted two places to the right.

Example

A double wound transformer supplied with 16 A at 180 V supplies a load of 36 A at a terminal voltage of 75.2 V. What is the efficiency of the transformer?

$$\text{Efficiency} = \frac{\text{power out}}{\text{power in}}$$

$$= \frac{36 \times 75.2}{16 \times 180}$$

$$= \frac{2707.2}{2880}$$

$$= 0.94 \text{ p.u.}$$

or alternatively:

$$\% \text{ Efficiency} = \frac{\text{power out} \times 100}{\text{power in}}$$

$$= \frac{270720}{2880} = 94\%$$

There is very little advantage to the second method as it has to be converted back to a per unit value before it can be used in any subsequent calculations.

Now work through the following example to see how transformer efficiency is incorporated into calculations of this type.

Example

A 150 V 10 kVA single phase transformer has a full-load efficiency of 96%. If the primary voltage is 250 V determine the

(a) transformer ratio

(b) full-load primary current

Solution

(a) Transformer ratio

$$= \frac{U_1}{U_2}$$

$$= \frac{250}{150}$$

$$= 1.66 : 1$$

(b)

$$\text{efficiency} = \frac{\text{output}}{\text{input}}$$

$$\text{input} = \frac{\text{output}}{\text{efficiency}}$$

$$= \frac{10000}{0.96}$$

$$= 10416.66 \text{ VA}$$

At a primary voltage of 250 V this would mean that the transformer takes a current of

$$I_1 = \frac{10416.66}{250}$$

$$= 41.66 \text{ A}$$

Try this

A double wound transformer having a full-load efficiency of 98% is supplied with 24 A at 200 V. If its secondary voltage is 120 V, what is the secondary current at full load?

Transformer losses

A transformer is not 100% efficient because it has losses.

If one tenth of the power delivered to the transformer is taken up by the losses then only nine tenths of the input power would end up as output.

In other words

Input = Output + Losses

Transformer losses are generally grouped into two different categories.

- copper losses and
- iron losses

Copper losses

These are quite simply explained as the power lost through heat in a loaded winding.

Power is lost in a winding as a result of current passing through a resistance. The resistance of the winding will be determined by the length and cross-sectional area of the conductor. Nothing much can be done about the length because a transformer winding will need to have a pre-determined number of turns in order to meet its design specification. If however the cross sectional area is too small, then the efficiency of the transformer will suffer due to excessively high copper loss.

In terms of current and resistance we can express power loss as follows:

Power $= I^2 R$ Watts

From this it can be seen that power loss is proportional to the SQUARE of the current in the windings.

This is an important factor in transformer operation as any reduction in load current has a significant effect on copper loss, equivalent to the square of the fraction of full load current.

In other words:
A transformer at half full-load will have copper losses equivalent to one quarter of the full-load copper losses.

At one quarter of full-load, the copper losses will be one sixteenth of what they would have been if the transformer had been fully loaded.

By comparison, a transformer running at 10% overload would show a copper loss of 21% more than the design full-load value and consequently its efficiency would be adversely affected.

Example

A transformer has a full-load copper loss of 200 Watts. What is the copper loss at
(a) two-thirds full load?
(b) one third full load?

Solution

(a)
Copper loss = f.l. copper loss × the square of the fraction of full load

= $200 \times (0.666)^2$ = 88.7 W

b)
Copper loss = $200 \times (0.333)^2$ = 22.18 W

The short circuit test

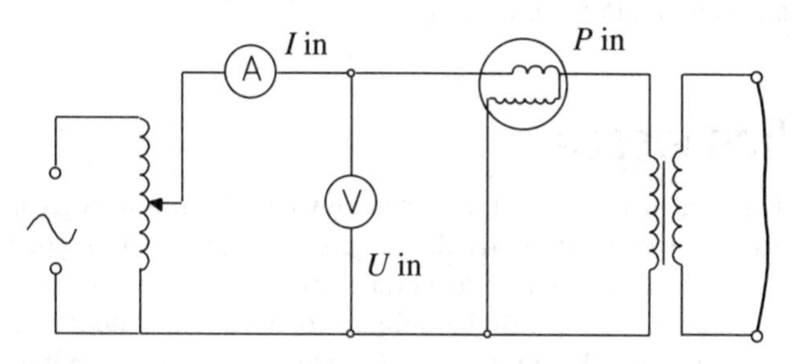

Figure 5.5

The transformer secondary in Figure 5.5 is short circuited and the primary is connected to a variable voltage a.c. source.

The primary voltage is increased until full load primary current is registered on the ammeter.

Since the primary voltage is at a very low level and the flux produced by one winding is virtually eliminated by the flux from the other, iron losses will be negligible.

This means that the power indicated by the wattmeter will represent the full-load copper loss only.

Example

A 300/75 V transformer has a full load primary current of 12.5 A.

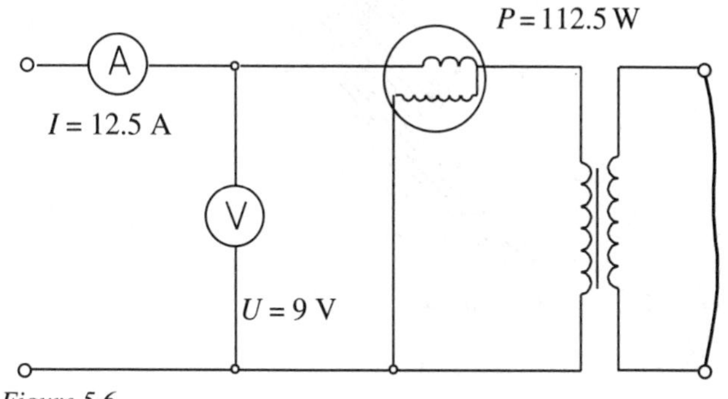

Figure 5.6

A short-circuit test gives the following information.

Short circuit Voltage = 9 V

Short circuit Current = 12.5 A

Short circuit Power = 112.5 W

For the transformer in this example the copper loss would account for,

$$= \frac{\text{short circuit power}}{\text{full load power}}$$

$$= \frac{112.5}{3750}$$

$$= 0.03 \quad \text{or 3\% of full load power}$$

This is of course the copper loss only and we must determine the iron loss in order to complete the picture.

Iron losses

The core material of a transformer must be chosen with great care in order to minimize the effects of iron loss. From your study of electromagnetic materials you will have learned that the core material will be subject to a certain amount of hysteresis loss (For revision see Chapter 1). This is the amount of energy input required to magnetise the core during the positive half cycle of the supply voltage then to de-magnetise and re-magnetise to the opposite polarity during the negative half cycle.

The hysteresis loss is proportional to the area enclosed by the loop and consequently an efficient transformer core material is one which is easily magnetised and de-magnetised and forms a narrow loop thus enclosing as small an area as possible. It must be remembered that the hysteresis loss is not a "one off" feature but is repeated with every cycle of the supply current.

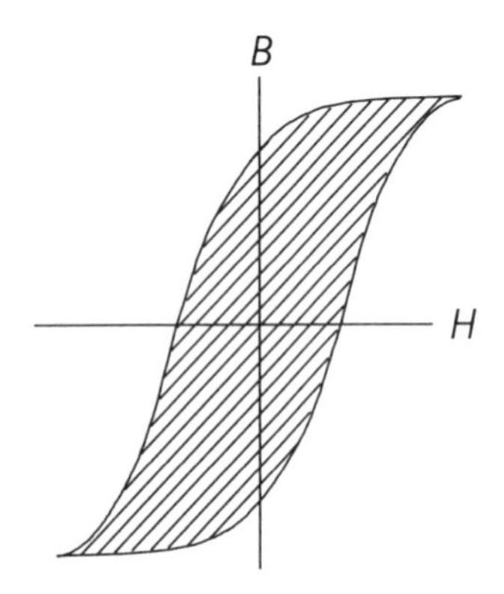

Figure 5.7 *High loss*

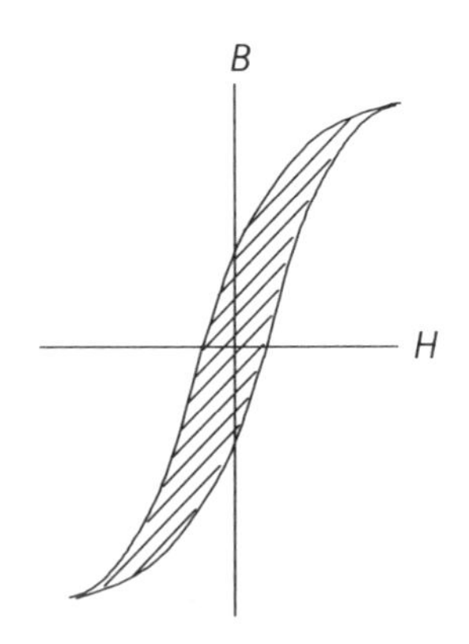

Figure 5.8 *Low loss*

Eddy current loss

Because of the nature of induced currents it is inevitable that alternating current will be induced in any conductor which is situated in an alternating magnetic field.

Since the core is likely to be made of ferromagnetic (iron based) material this is in itself a conductor material. Locally circulating (eddy) currents will be induced in the core material and unless prevented from doing so, these will circulate in the core performing no useful function and dissipate their energy in the form of unwanted heat thus reducing the overall efficiency of the device.

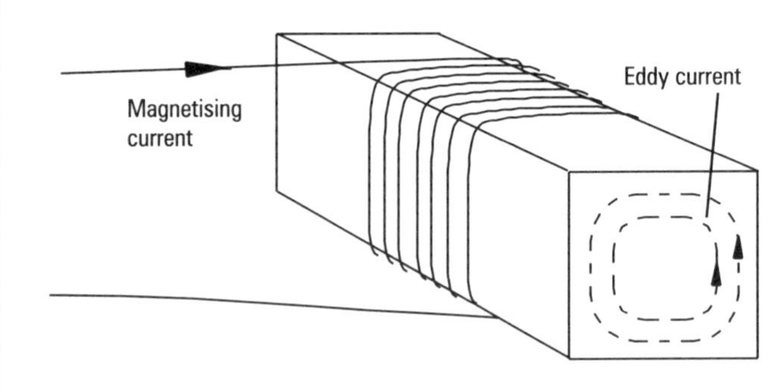

Figure 5.9

In a solid core, eddy currents will be induced and will flow freely in the core material.

If the core is made up of thin sheets or laminations as they are called, and each lamination is insulated from the next by a microscopically thin layer of insulating material then the eddy currents are virtually eliminated. The overall size of the core is only marginally greater than it would have been as a solid core and the flux path is not impeded.

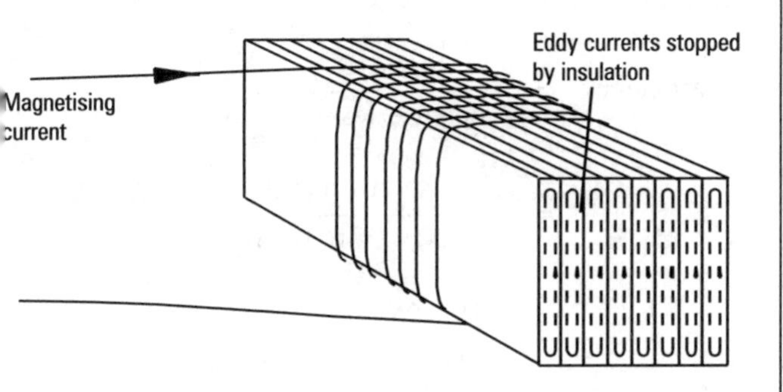

Figure 5.10

Open circuit test

Taking the same transformer as used in the short circuit test, with the secondary winding open circuited, the full supply voltage is applied to the primary.

In this test it will be found that the open circuit current is very small in comparison to the full-load current.

Because this represents a very small fraction of the full-load current which is then squared, the effects of copper loss will be so small as to have no significance to the results obtained.

Example

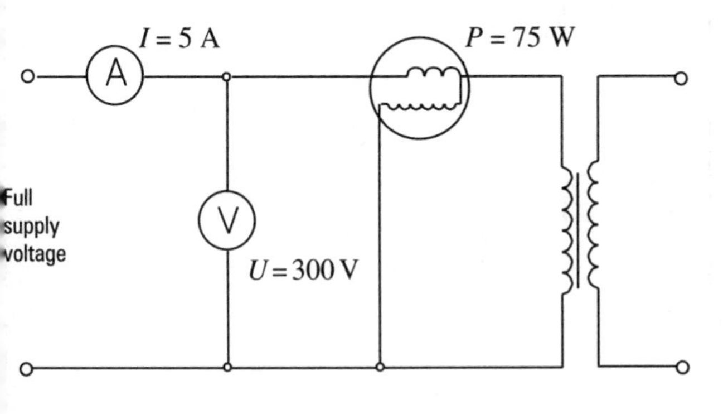

Figure 5.11

Open circuit Voltage = 300 V

Open circuit Current = 5 A

Open circuit Power = 75 W

Because the core is now fully magnetised at its working frequency and the copper loss has been eliminated, the whole of the power consumed by the transformer is being taken up by the iron losses.

Therefore

$$\text{iron loss} = 75\text{ W}$$

As a proportion of full-load power this would be

$$\frac{\text{iron loss}}{\text{total power}} = \frac{75}{3750} = 0.02$$

$$= 2\% \text{ of the total}$$

Given that we have iron losses of	0.02
and copper losses of	0.03
then the total losses are	0.05

or 5% of the power input .

It is normally assumed that iron losses will be constant under all load conditions as variations will be negligible.

$$\text{efficiency} = \frac{\text{output}}{\text{input}}$$

$$= \frac{\text{input} - \text{losses}}{\text{input}}$$

$$= \frac{3750 - (112.5 + 75)}{3750}$$

$$= 0.95$$

or 95% efficient at full load

It is important to stress that this efficiency will only exist at full load as the copper losses will vary according to load conditions.

Now work through the following example.

Example

A 400/230 V transformer has a rated output of 6.3 kW and tests have shown that at full load it has an iron loss of 800 W and a copper loss of 1200 W. Determine the efficiency of this transformer, at

(a) full load
(b) half load

Solution

(a) At full load,

$$\text{efficiency} = \frac{\text{output}}{\text{output} + \text{losses}}$$

$$= \frac{6300}{6300 + 800 + 1200}$$

$$= \frac{6300}{8300}$$

$$= 0.759\ (75.9\%)$$

b) At half load

Iron losses $=$ 800 W

Copper loss $= 1200 \times 0.5^2 = 300$ W

Total losses $=$ 1100 W

$$\text{Efficiency} = \frac{\text{half output}}{\text{half output} + \text{losses}}$$

$$= \frac{3150}{3150 + 1100}$$

$$= 0.741 \qquad (74.1\%)$$

As you will have seen, the efficiency is less at half load than it was at full load even though the copper loss has been reduced to a quarter of its full load value. This is because the iron loss has remained constant and now represents a greater proportion of the transformer input.

Maximum efficiency

Maximum efficiency occurs when the iron losses and copper losses are equal.

In the previous example this would be when the copper loss is 800 Watts.

This represents two thirds of the full load copper loss but it does not occur at two thirds of the transformer full load.

Since copper loss is proportional to the square of the fraction of full load, then the transformer must be loaded to a part of the full load which is equivalent to the square root of this fraction.

i.e.

The proportion of full load which gives maximum efficiency

$$= \sqrt{\frac{\text{iron loss}}{\text{full load copper loss}}}$$

From the previous example:

Transformer loading for maximum efficiency

$$= \sqrt{\frac{800}{1200}}$$

$$= 0.8165$$

Therefore maximum efficiency will occur when this transformer is loaded to 0.8165 (81.65%) of its rated full load.

The actual efficiency will be

$$\text{efficiency} = \frac{\text{output}}{\text{output} + \text{losses}}$$

$$= \frac{6300 \times 0.8165}{(6300 \times 0.8165) + 1600}$$

$$= \frac{5143.95}{6743.95}$$

$$= 0.7627 \qquad (76.27\%)$$

Which you will see, is a higher efficiency than either of the two previous examples.

Try this

A transformer has a rated output of 3.6 kW. It has full load copper losses of 800 W and iron losses of 400 W. Determine the

(a) efficiency of the transformer at full load

(b) efficiency at half load

(c) proportion of full load at which efficiency is a maximum

Regulation

A transformer has losses which are affected by the load condition so it is inevitable that there must be some fall off in output voltage as the load increases.

This situation is very similar to the voltage drop which occurs in a loaded conductor. The voltage drop will vary according to the load current due to the fact that current is passing through a conductor which has resistance and the voltage drop is just another example of Ohm's Law in action.

The same effect is to be found in transformers and the voltage drop can be determined by measuring the output voltage on open circuit then measuring again when the transformer is loaded. Transformer regulation in its simplest form is the difference in voltage between no-load and full-load expressed as a percentage or per unit of the no-load voltage.

Percentage regulation

$$= \frac{(\text{no load Volts} - \text{full load Volts} \times 100}{\text{no load Volts}}$$

Example

A transformer has a no-load voltage of 224 V and a full load voltage of 116 V. What is the percentage regulation?

$$\% \text{ Regulation} = \frac{(224 - 216) \times 100}{224}$$

$$= 3.57\%$$

That was simple enough.

Example

A transformer with a percentage regulation of 2.6% has a full load output voltage of 194.8 V. From this information determine the open circuit voltage of the device.

Since the transformer has a percentage regulation of 2.6% then; 194.8 will represent 97.4% of no load voltage therefore:

$$\text{no load volts} = \frac{\text{full load volts}}{0.974}$$

$$= \frac{194.8}{0.974}$$

$$= 200 \text{ V}$$

Try this

1. A transformer has an open circuit voltage of 380 V which falls to 372 V when fully loaded. Calculate the percentage regulation.

2. A transformer having a percentage regulation of 3.4% has a full load output voltage of 400.9 V. What is its no load voltage?

Transformer impedance

The regulation of the transformer is obviously dependent on its impedance. Regulation percentages are normally quite low in large power transformers and this can be taken as an indication that the impedance of the device is low thus allowing a good transfer of energy with minimal losses.

It is quite simple to determine the impedance of a transformer from the regulation as follows:

$$\text{Transformer impedance} = \frac{\text{voltage difference}}{\text{full load current}}$$

This is of course Ohm's Law as applied to a.c. circuits.

Example

A transformer has an open circuit voltage of 500 V which drops to 497 V at a load current of 125 A.

What is the impedance of the transformer?

$$Z_t = \frac{500 - 497}{125}$$

$$= 0.024\ \Omega$$

This would appear to be an acceptable state of affairs with regard to transformer efficiency and satisfactory operation.

What then would be the effect of a short circuit fault close to the output terminals of the transformer?

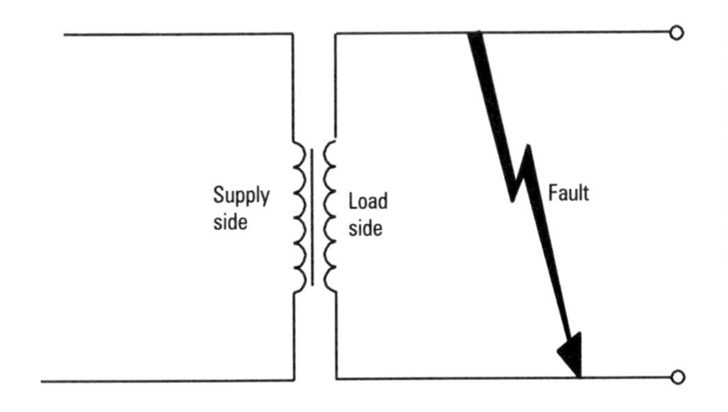

Figure 5.12

If this should occur, the only thing left to control the flow of current from the transformer into the fault would be the transformer's impedance.

$$I_f = \frac{U_o}{Z_t}$$

$$= \frac{500}{0.024}$$

$$= 20.83 \text{ kA (20833 Amperes)}$$

This is by any standards a very large current which could cause considerable damage to cables and plant if not quickly dealt with.

Try this

A transformer has a no load output voltage of 440 V which falls to 435 V when loaded with 100 A. From this information determine the current which would flow in a short circuit of negligible impedance close to the transformer.

Percentage impedance

It is common practice when calculating fault currents in a transformer to use the percentage impedance. This is a percentage value which can be derived from the transformer regulation as follows:

Percentage impedance = Percentage regulation

The power developed in a short circuit fault can then be determined by:

$$\text{Short circuit kVA} = \frac{\text{Rated kVA} \times 100}{\text{\% impedance}}$$

From the previous Try this:

$$\text{\% impedance} = \frac{5 \times 100}{440}$$

$$= 1.136\%$$

Rated kVA = Rated volts × Rated current
= 440 × 100
= 44 kVA

$$\text{Short circuit kVA} = \frac{\text{Rated kVA} \times 100}{\text{\% impedance}}$$

$$= 3873.2 \text{ kVA}$$

$$\text{Short circuit current} = \frac{\text{Short circuit kVA}}{\text{Voltage}}$$

$$= 8.8 \text{ kA (8800 A)}$$

Short circuit currents in installation faults

To determine the level of fault current in an installation it is necessary to consider the whole of the fault loop. This starts with the transformer as the source of the fault current but includes the impedance of the main's network and the resistance of the installation cables up to the point of the fault.

Example

The installation is supplied by a 500 kVA transformer having a resistance of 0.04 Ω and a reactance of 0.3 Ω. The mains cable consists of 100 metres of 70 mm^2 aluminium cable with a resistance of 0.105 Ω and a reactance of 0.017 Ω. The fault is 10 metres from the source in a 16 mm^2 twin cable with a resistance of 0.028 Ω.

Calculate the current in the fault.

The total impedance of the loop is made up as follows:-

Transformer	R_t	= 0.04	X_t	= 0.3
Main	R_m	= 0.105	X_m	= 0.017
Installation	R_{inst}	= 0.028		
Total	R	= 0.173	X	= 0.317

giving a total impedance of 0.36 Ω$(Z = \sqrt{R^2 + X^2})$

At a voltage of 230 V this would produce a current of

$$\frac{230}{0361} = 637.12 \text{ A}$$

Try this

Calculate the current in a short circuit fault in 20 metres of 35 mm^2 twin cable fed from a 95 mm^2 main 60 metres long which originates in a 400 V transformer having a resistance of 0.04 Ω and a reactance of 0.25 Ω. The main cable has a total resistance of 0.03 Ω and a reactance of 0.009 Ω and the total resistance of the installation cable is 0.025 Ω, with a reactance of 0.03 Ω.

Transformer types

Auto-transformer

Not all transformers are of the double wound type. The auto transformer has only one winding and is capable of stepping up and down the voltage as effectively as the double wound variety.

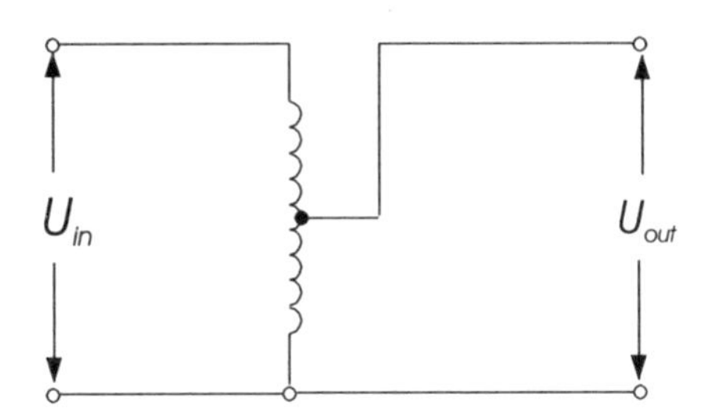

Figure 5.13 *Step down auto-transformer*

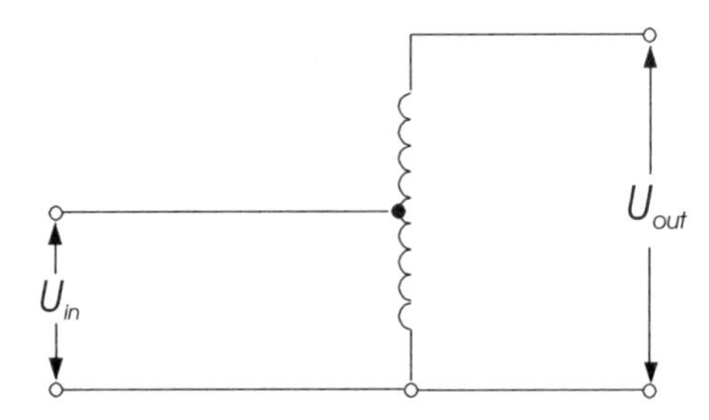

Figure 5.14 *Step up auto-transformer*

One important advantage of the autotransformer is that the part of the winding which is common to both primary and secondary current carries the difference between the two currents (Figure 5.15).

Where this type of transformer is used to make small adjustments in mains voltage, the currents are very nearly equal and therefore the resultant current in the common part of the winding is quite small. For this reason, the common section can be wound in comparatively light wire resulting in considerable cost and weight savings as compared with double wound types.

The main problem with the auto transformer arises from the fact that it does not have an isolated secondary. Another unfortunate feature is that if the common terminal should become disconnected then full input voltage will appear at the output terminals.

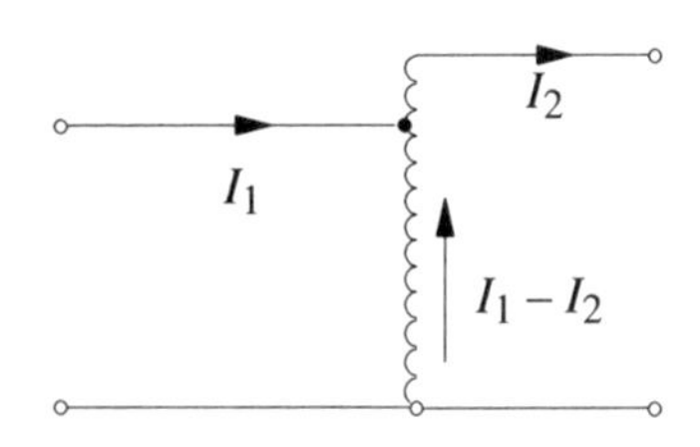

Figure 5.15

Transformer windings

It is quite convenient to draw a transformer as two windings wound on separate limbs of a common core. Although this device would work, it is not the most efficient form of construction. Better flux linkage is achieved by building up the windings in alternate layers or discs so that the flux produced by one winding is more readily adjacent to the turns of the other.

Concentric windings are used for all sizes of power transformers. This technique is used to build up a cylindrical coil assembly consisting of alternating layers of primary and secondary turns which are insulated from each other by interleaved insulation to prevent breakdown.

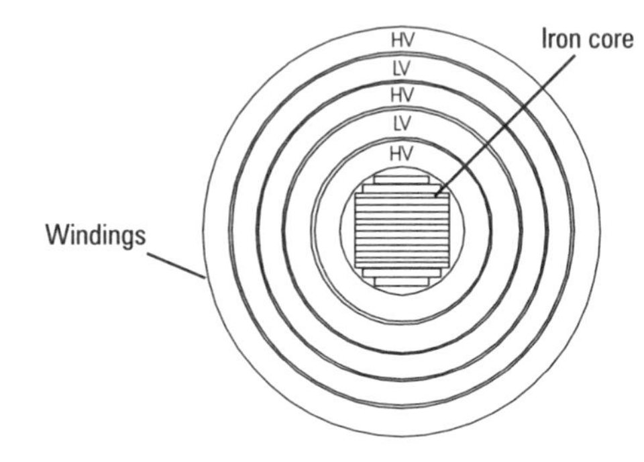

Figure 5.16 *Concentric windings*

Disc or sandwich winding involves assembling and interconnecting alternate sections of the winding by placing one on top of the other in a stacking arrangement.

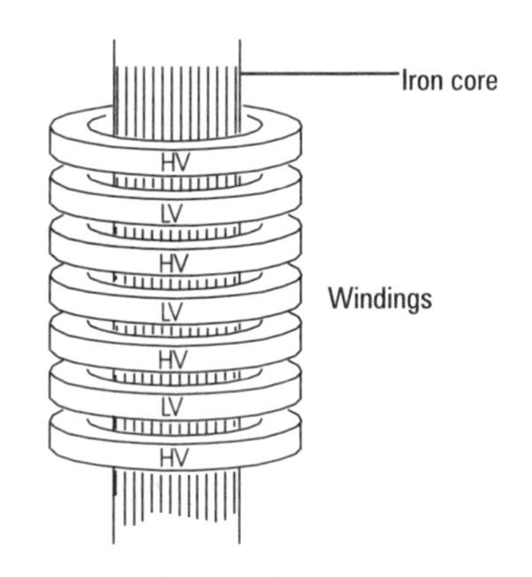

Figure 5.17 *Disc or sandwich windings*

The core is built up around the coils using laminations which have been stamped out to give the correct shape when fully assembled.

Small power transformers are usually of the "core" or "shell" type assembly.

A core type will usually have its windings evenly distributed over the two outer limbs of a rectangular core with a single "window" in the centre.

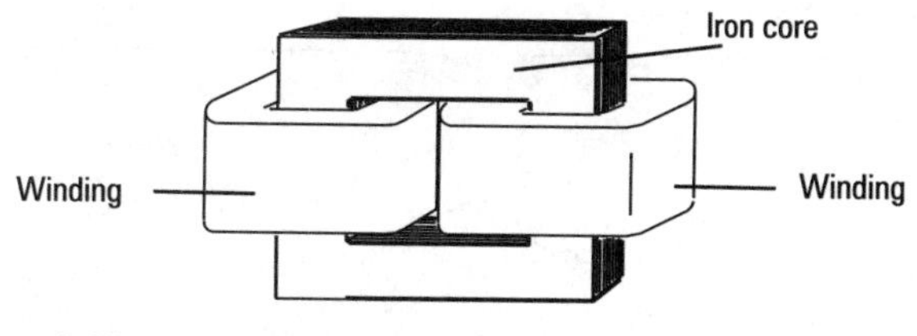

Figure 5.18 *Core type transformer*

The shell assembly uses a three-limb construction with both windings mounted on the centre limb.

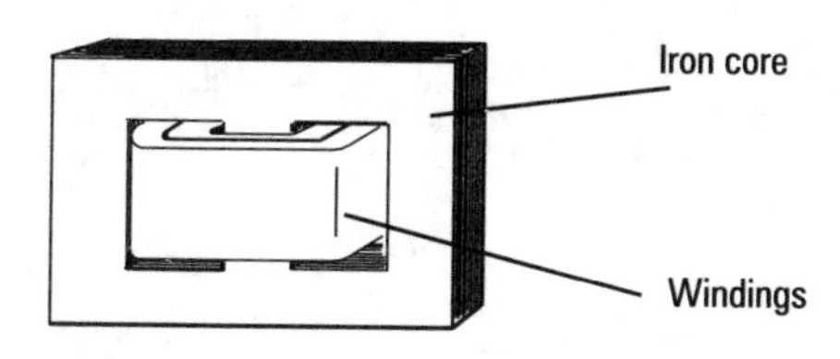

Figure 5.19 *Shell type transformer*

Three phase transformers

Three phase transformers are an essential feature of transmission and distribution systems. They are to be found in power stations, grid switching stations and sub-stations all over the country wherever the supply voltage changes from one level to another.

The basic construction consists of three pairs of windings; one High Voltage, the other Low Voltage, mounted on separate limbs of a three-limb core. The coils are wound and distributed in the same manner as single phase windings but the level of insulation between windings will take into account the high voltages likely to be present.

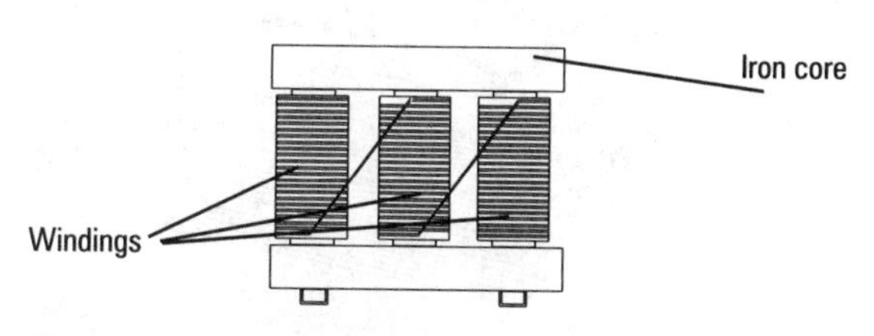

Figure 5.20 *Three phase transformer*

The transformer ratio rules will still apply but with a three-phase transformer there is always the choice of STAR or DELTA connection.

The effect of this can be seen as follows.

Example

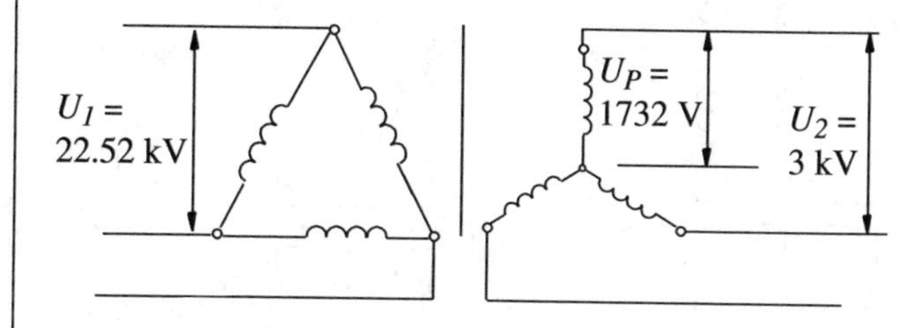

Figure 5.21

The transformer has a ratio of 13:1 and is supplied at 22.52 kV.

The primary windings are DELTA connected and the LV side is connected in STAR.

The voltage across each of the HV windings is 22.52 kV and assuming negligible losses, each LV winding will have a voltage of

$$\frac{22.52 \times 10^3}{13}$$

i.e. 1732.3 V

Since these windings are now to be connected in star connection the terminal voltage will be

$$U_2 \times \sqrt{3} \quad = \quad 1732.3 \times \sqrt{3} \quad = 3 \text{ kV}$$

Try this

T_1 has a ratio of 6.9:1 and is connected Star/Delta to a 33 kV three phase supply.
T_2 has a ratio of 47.8:1 and is connected Delta/Star to the output circuit from T_1.
What is the voltage at the LV terminals of T_2?

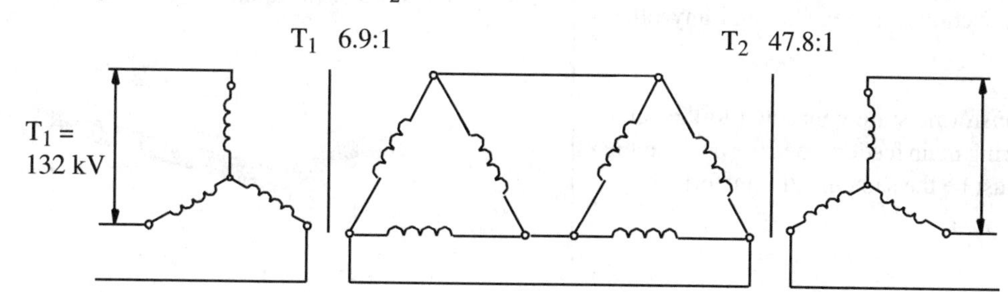

Terminal markings

Transformer terminals are generally marked with CAPITAL LETTERS, for example, A1 B1 C1 on the HV side and lowercase letters d2 e2 f2 on the LV side.

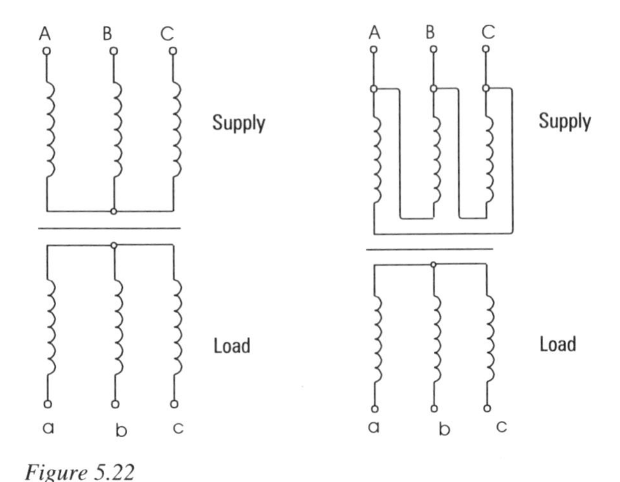

Figure 5.22

The windings themselves may be connected in star or delta as required and the transformer may carry an additional marking to indicate that this has been done.

A transformer which is marked D.y is delta connected on the HV side (capital D) and star connected on the LV (lower case y).

Other variations would be y.D; D.d; Y.d or Y.y.

In addition to this, a number may be included in the code to indicate the phase relationship between the windings. This number is based on the clock face and uses the numbers on the dial rather than the phase angle.

For example, a transformer marked Y.y 6 is star connected on both HV and LV windings but the LV is 180° out of phase with the HV.

Another may be marked Y.d 8 to indicate that the delta connected LV winding is 120° out of phase with the star connected HV side.

Where a transformer is working independently of others, the phase angle of the output voltage creates no particular problem as there is no inter-connection between this and any other circuits.

Where two or more transformers are connected to the same circuit, as in the case of ring main feeders, then the phase angle of the output voltage must be the same in all transformers.

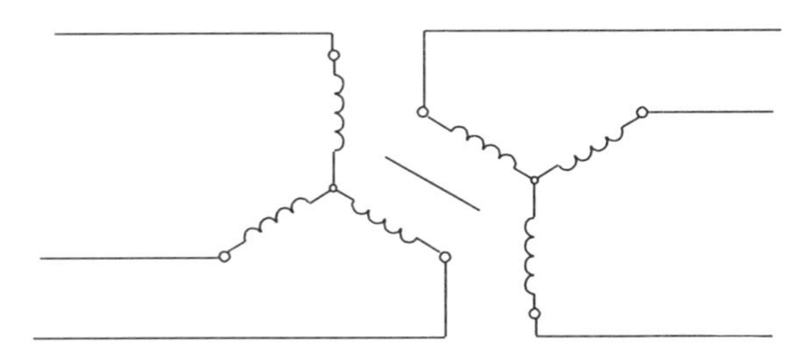

Figure 5.23 *Y.y 6 connected transformer*

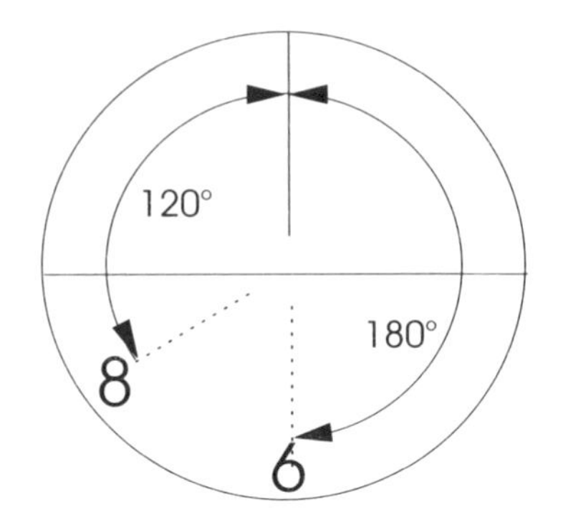

Figure 5.24

Construction and enclosures

Small power transformers for use in equipment are usually of open construction, with varnished windings, and are air cooled.

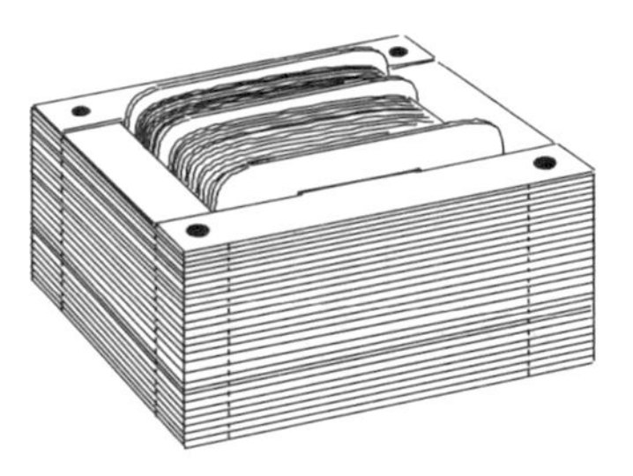

Figure 5.25 *Small air cooled power transformer*

Larger transformers for medium power applications may be housed in metal tanks which are then filled with mineral oil which helps to insulate the windings. Excess heat is dissipated through the sides of the tank.

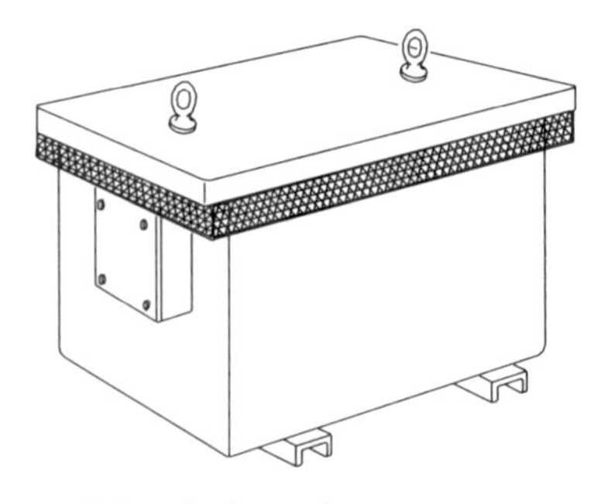

Figure 5.26 *Oil cooled transformer*

Large power transformers are enclosed in a metal tank with cooling tubes fitted to the outside through which the oil is free to circulate. The oil serves as an insulating and a cooling medium as the natural convection of the liquid carries away the heat from the windings as well as insulating one from the other.

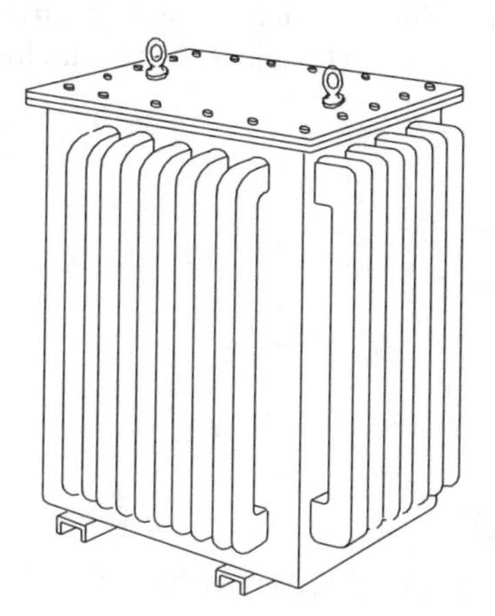

Figure 5.27 *Large oil filled transformer*

A gas-pressure relay (Buchholz Relay) is used to protect large oil cooled transformers from explosion under fault conditions. Figure 5.28 shows a typical type that is fitted between the transformer and expansion tank.

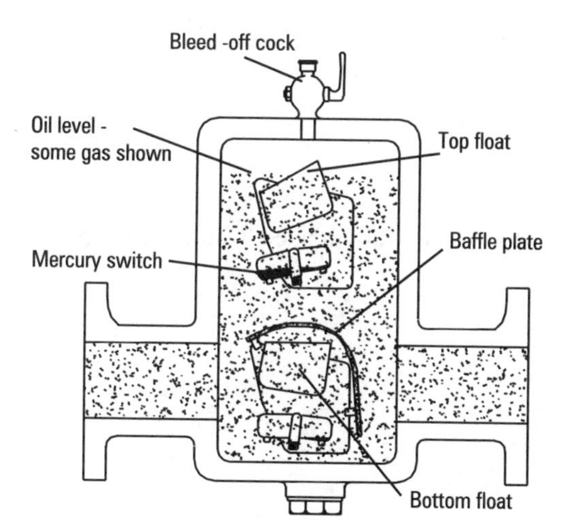

Figure 5.28 *Buchholz Relay*

When a fault develops in the transformer, gas is given off and this builds up in the top of the relay unit. The top float then rotates and sets off an alarm. If there is a serious fault, the pressure pushes on the baffle plates on the bottom float and the relay switches off the transformer.

Any large item of oil-filled plant such as a transformer or circuit breaker must be mounted over a pit or surrounded by a low wall which will contain all the oil in the event of leakage.

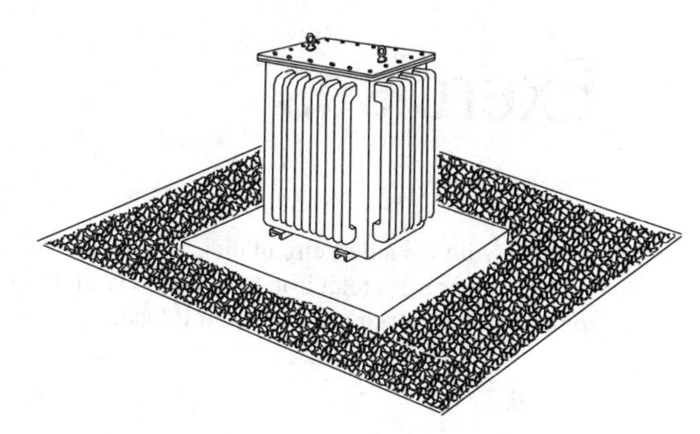

Figure 5.29 *Transformer mounted over a pit*

In the case of the pit, this is usually filled with loose stones or gravel to prevent anyone from falling in but will still hold all the oil contained in the tank or tanks if a leak should happen.

Exercises

1. (a) Explain, with the aid of circuit diagrams, how to determine the iron and full-load copper losses in a transformer.
   (b) A transformer has a total iron loss of 1.8 kW and full load copper losses of 4.1 kW. Calculate the total losses when the transformer is working at 80% of full load.

2. (a) Explain what is meant by copper losses and iron losses in a transformer.
   (b) A single-phase transformer rated at 35 kVA has an open circuit voltage of 240 V and losses of 750 W. The losses when the transformer was short circuited at full load were 850 W. Calculate the efficiency of the transformer at unity power factor at
   i) full load
   ii) 50% of full load

3. (a) Explain what is meant by the regulation of a transformer.
   (b) A transformer has an open circuit voltage of 420 V which falls to 415 V when fully loaded. Calculate the percentage regulation.

4. (a) Explain, with the aid of diagrams, what is meant by
   i) isolating transformers
   ii) auto transformers
   (b) A transformer on full load has losses of 8%. If the input is 35kVA calculate the output kVAr when the power factor is 0.8.

6

Measurement

Complete the following to remind yourself of some important facts on this subject that you should remember from the previous chapter.

1. Ignoring losses, what is the input current of a 230:110 V transformer when delivering 16 A at 110 V?

2. The transformer efficiency is at a maximum when?

3. What is the essential difference between an auto-transformer and a double wound transformer?

On completion of this chapter you should be able to:

- describe the basic construction of given analogue instruments
- calculate the resistances required to extend the range of instruments
- describe how the resistance of components can be measured
- describe methods of measuring temperature and speed
- describe the measurement of power and power factor
- describe the connection of current and voltage transformers
- identify instruments suitable for "on-site" measurements
- calculate the power in a three-phase circuit, given wattmeter readings

Measuring instruments

Because electrical quantities have no obvious physical properties, the only way to obtain a reliable indication of their presence and magnitude is by measuring the effects that they produce.

Electrical measuring instruments use a current, or the presence of an electrical charge, to create a physical change which can be translated into some form of visible output.

Measuring instruments have to be accurate and reliable if they are to be of any use. The accuracy and reliability of an instrument will depend on the soundness of its design, its construction and the choice of components used in its construction.

It is not the purpose of this section to consider the construction of such instruments but some attention will be given to their operating principles so that you can have a better understanding of the techniques involved in their application.

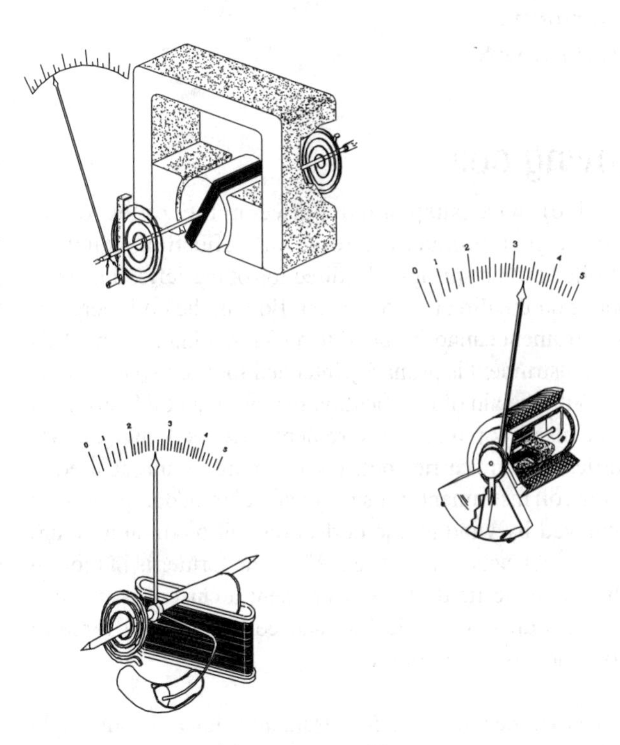

Figure 6.1 Measuring instruments

Analogue instruments

In the majority of cases an analogue instrument takes an electric current and uses this to produce a mechanical deflection. The deflecting force can be the force on a conductor in a magnetic field, as in the case of the moving coil instrument, or the force exerted between the poles of a magnet as in the moving iron. In either case, an electromagnetic device is used to create the deflecting force. The force will depend on the current flowing in the instrument. The greater the current - the greater the force.

In order to measure the strength of the deflection, it must be restrained by a controlling force. The controlling force is usually provided by two spiral springs which hold the movement at rest when the amount of torque exerted by the deflecting force is matched with the controlling torque of the springs. When there is no current in the measuring instrument the control springs will hold the movement at zero.

In addition to deflection and controlling forces, it is necessary for some form of damping to be provided for analogue instruments. Without a damping mechanism, a meter movement will oscillate to and fro for a considerable time before coming to rest at its true deflection. This makes it very difficult to obtain an accurate reading, more especially so if the deflecting current changes before the movement has come to rest.

The various types of analogue measuring instrument can be broken down into five groups.

- Moving coil
- Moving iron
- Electrodynamic
- Electrostatic
- Thermocouple

Moving coil

A coil of wire suspended between the poles of an electromagnet produces a force of deflection which is controlled by hair springs. The direction of the deflecting force depends on the direction of current flow in the coil therefore this instrument cannot respond to an alternating current. This type of instrument is primarily intended for use in d.c. circuits only. With the aid of rectification the moving coil instrument can be used for a.c. measurement provided that certain modifications are carried out. The damping technique used in moving coil instruments is usually eddy current damping. This is achieved by winding the deflection coil on an aluminium former which acts as a damper. When the former is in motion in the magnetic field of the instrument a current is induced which sets up a magnetic flux and consequently a force to oppose the movement of the coil.

The scale of the moving coil instrument is linear i.e. the scale divisions are the same size from zero to full scale deflection.

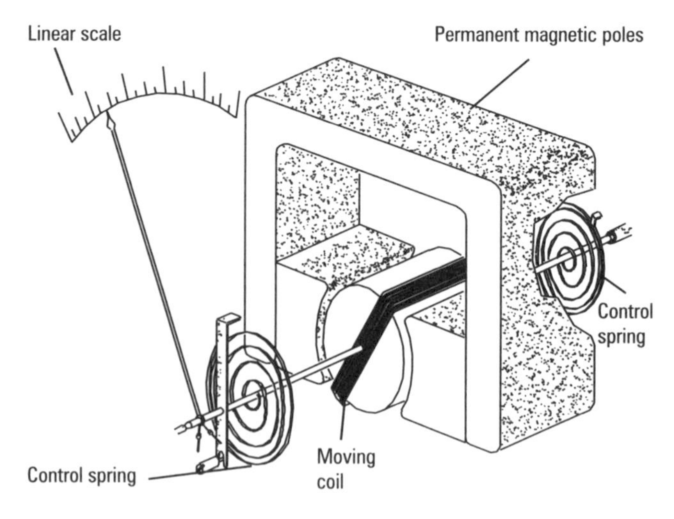

Figure 6.2 *Moving coil meter*

Moving iron

Moving iron instruments work on the principle of magnetic attraction or repulsion depending on their design. They normally use a coil spring controlling mechanism but some may use a simple gravity device. The moving iron instrument is deflected by direct current or r.m.s. alternating current therefore it does not need a rectifier when connected to a.c. circuits.

Without damping, a moving iron instrument swings quite appreciably before coming to rest in its deflected state. The damping mechanism most commonly used is in the form of an air vane or damper which is enclosed in a chamber. The air pressure built up by the damper acts as a decelerating force in either direction and quickly brings the movement to rest. The scale of a moving iron instrument is non-linear as the scale divisions become smaller as the deflection approaches full scale.

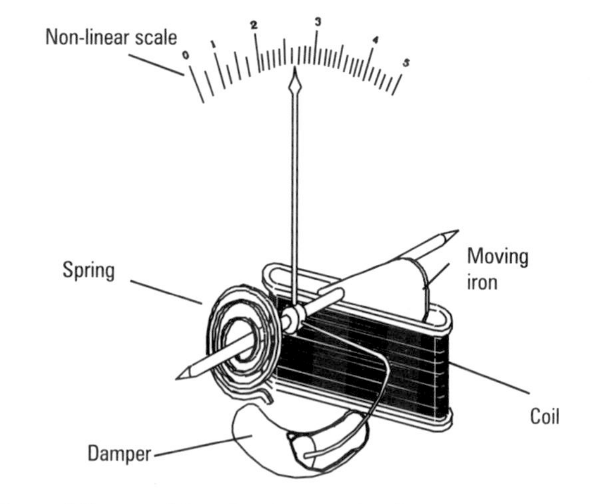

Figure 6.3 *Attraction type moving iron meter*

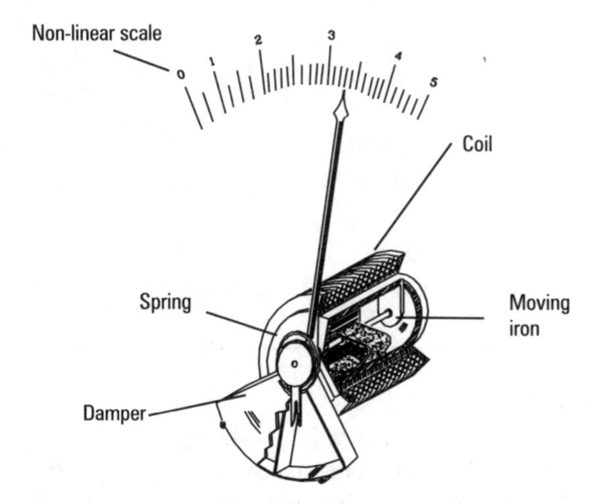

Figure 6.4 *Repulsion type moving iron meter*

Electrodynamic

This type of moving coil instrument uses an electromagnet, instead of a permanent magnet, to produce the magnetic field which surrounds the coil. Although electrodynamic ammeters and voltmeters can be found, the main application for this type of instrument is as a wattmeter.

The control of the movement is by hairsprings. As an electrodynamic or dynamometer wattmeter the deflection is produced between the fixed current coil and the moving voltage coil acting at the same instant giving a scale reading which is proportional to the true power in the measured circuit.

Air damping is usually provided for this type of instrument and the scale is linear.

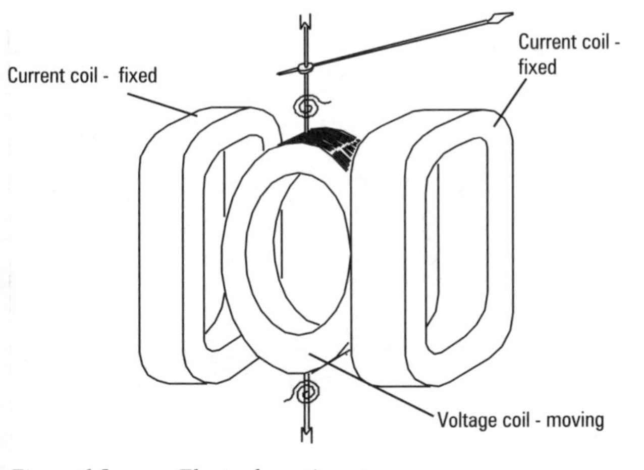

Figure 6.5 *Electrodynamic meter*

Electrostatic

This is an instrument of some academic interest as it uses the principle of electrostatic attraction rather than electromagnetism as found in most other types.

It consists of two sets of vanes, one fixed, one moving which are held separate from each other by the control springs. When connected across an a.c. or d.c. voltage a force of attraction proportional to the square of the voltage pulls the moving vanes into the gaps between the fixed vanes.

Electrostatic instruments are only really effective at high voltages but have the advantage that they do not draw any current from a d.c. source and when connected to an a.c. supply, the current drawn is negligible.

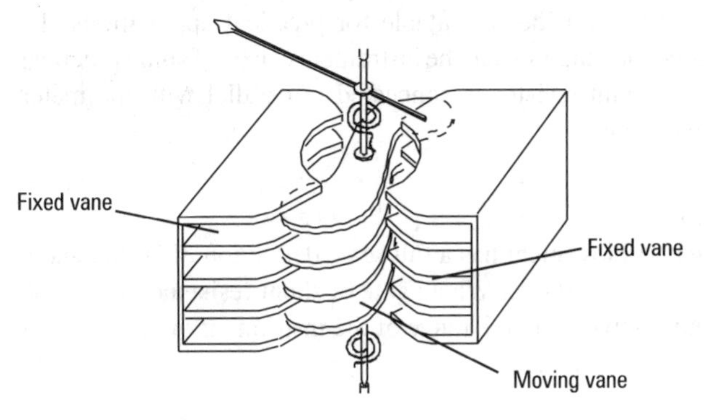

Figure 6.6 *Electrostatic instrument*

Thermocouple instruments

A thermocouple is a device consisting of wires of two different metals which are connected together to form junctions. When one of the junctions is heated, a potential difference will be created between it and the cold junction. The potential difference will be very small but nevertheless it will be proportional to the temperature difference between the junctions.

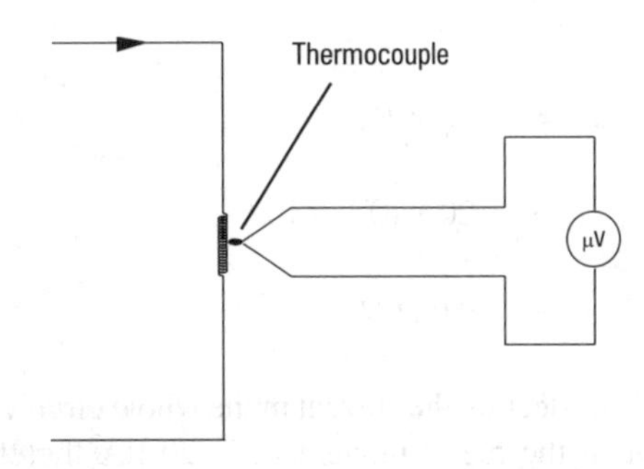

Figure 6.7 *Thermocouple instrument*

If, instead of measuring the current or voltage directly, the measured quantity is fed into a heating element, the heat produced will be determined by the measured current. A thermocouple embedded in the heating element will produce a potential difference proportional to the temperature, and this in turn will be used to deflect the measuring instrument. Examples of their use will be covered later in Chapter 7.

The main advantage of this type of instrument is that it can be used for d.c. and will give an accurate indication of the r.m.s. value of any a.c. source even at very high frequencies where direct reading electromagnetic instruments are no longer reliable due to their impedance.

Ammeters and voltmeters

When considering ammeters and voltmeters, one can say that there is no fundamental difference between them.

In the case of the ammeter, the basic movement is designed to give full scale deflection at a very low current, far lower than would be considered suitable for practical applications. To extend the range so that the instrument can be of some practical use, a shunt resistor is connected in parallel with the meter movement.

Example

A meter movement has a full scale deflection of 20 μA and a resistance of 100 Ω. What value of shunt resistance would be required to extend the range of the instrument to read 10 A at full scale deflection?

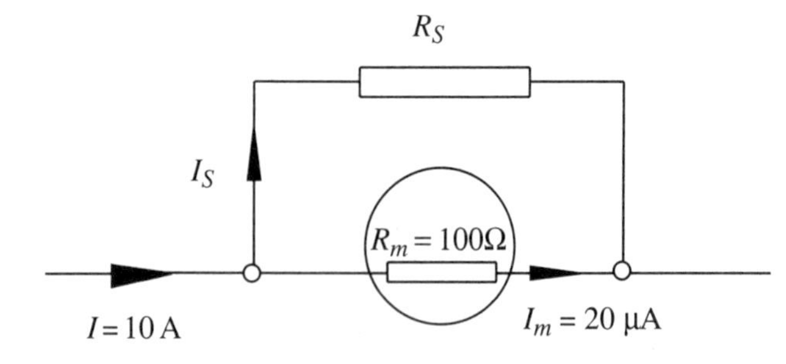

Figure 6.8

At full scale deflection, the voltage across the meter movement is

$$U_m = I_m \times R_m$$

$$= 20 \times 10^{-6} \times 100$$

$$= 0.002 \text{ V}$$

At full scale deflection the current in the whole circuit is 10 A, the current in the meter movement is 20 μA therefore the current in the shunt is (10 – 0.000002)

$$I_s = 9.99998 \text{ A}$$

$$U_s = I_m = 0.002 \text{ V}$$

(because they are in parallel)

$$R_s = \frac{U_s}{I_s}$$

$$= \frac{0.002}{9.99998}$$

$$= 0.0002\ \Omega$$

(2.000004×10^{-4} to be precise)

Try this

A meter movement has a full scale deflection of 1 mA and a resistance of 20 Ω. Determine the value of shunt resistance required to make it suitable for current measurement up to 25 A.

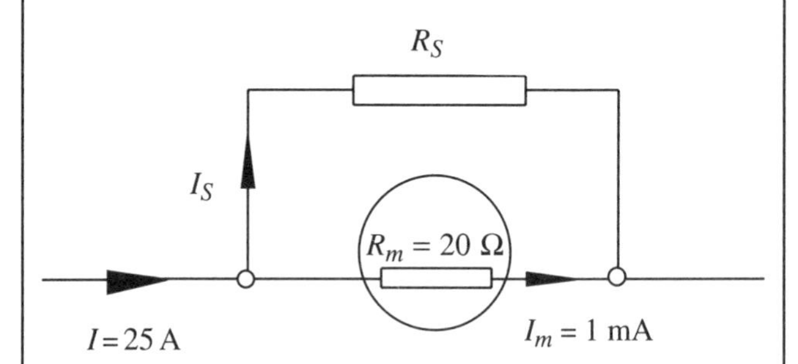

Multi-range instruments

With the aid of a selector switch and a number of resistors it is possible to connect one meter movement in such a way that it can be used to measure current on a number of ranges.

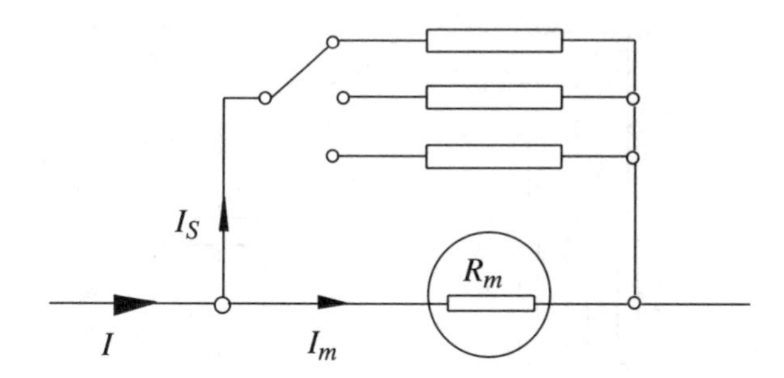

Figure 6.9

Example

A 0.1 mA movement with a resistance of 40 Ω can be used to measure current on the following scales:

0–1 A: 0–5 A: 0–10 A

I_m = 0.0001 A, and R_m = 40 Ω ∴ U_m = 0.004 V (4 mV)

Scale 1 (1 A)

$$I_s = I - I_m = 1 - 0.0001 = 0.9999\text{ A}$$

$$R_s = \frac{U_m}{I_s} = \frac{0.004}{0.9999} = 4.0004\text{ m}\Omega$$

(0.0040004 Ω)

Scale 2 (5 A)

$$I_s = I - I_m = 5 - 0.0001 = 4.9999\text{ A}$$

$$R_s = \frac{U_m}{I_s} = \frac{0.004}{4.9999} = 8.00016 \times 10^{-4}\,\Omega$$

(0.000800016 Ω)

Scale 3 (10 A)

$$I_s = I - I_m = 10 - 0.0001 = 9.9999\text{ A}$$

$$R_s = \frac{U_m}{I_s} = \frac{0.004}{9.9999} = 4.00004 \times 10^{-4}\,\Omega$$

(0.000400004 Ω)

Try this

Calculate the value of resistors so that a 1 mA meter movement with a resistance of 10 Ω can be used to measure currents in the following ranges:

(a) 0–5 A
(b) 0–10 A
(c) 0–15 A

Voltmeters

The same meter movement is used for voltage measurement but in the case of the voltmeter, the range is extended by connecting additional resistance in series with the meter movement.

Example

A basic meter movement has a full scale deflection current of 20 μA and a resistance of 100 Ω. If this is to be used to measure voltages up to 1 V, the full scale deflection current must flow through the instrument when a voltage of 1 V is connected across the circuit.

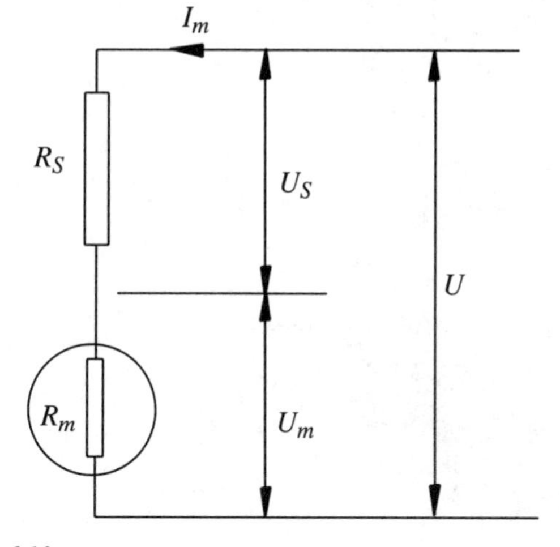

Figure 6.10

What has happened is that the bulk of the applied potential difference has been dropped across the series resistor leaving just the full scale deflection voltage across the meter movement.

$$U_m = I_m \times R_m$$
$$= 20 \times 10^{-6} \times 100$$
$$= 0.002 \text{ V}$$

$$U_s = U - U_m$$
$$= 1 - 0.002$$
$$= 0.998 \text{ V}$$

$$R_s = \frac{U_s}{I_m}$$

$$= \frac{0.998}{0.00002}$$

$$= 49900 \ \Omega \ (49.9 \text{ k}\Omega)$$

Try this

A meter movement with a full scale deflection current of 0.1 mA and a resistance of 10 Ω is to be adapted for use as a voltmeter with a range of 0–40 V. Determine the value of series resistor required for this purpose.

Multi-range voltmeter

Figures 6.11 and 6.12 show how a single meter movement, a selection of resistors and a rotary switch can be used to cover several ranges.

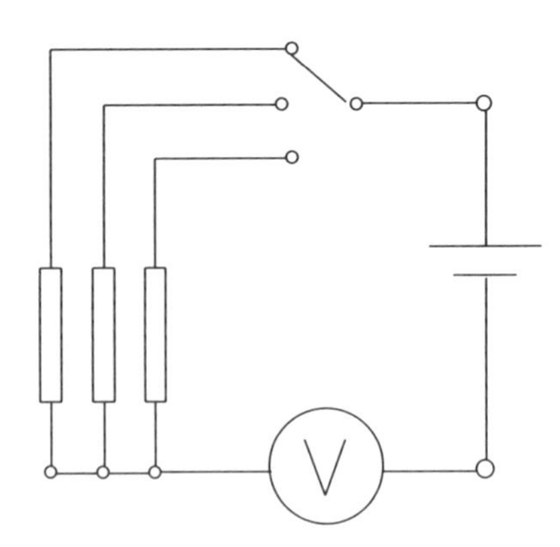

Figure 6.11

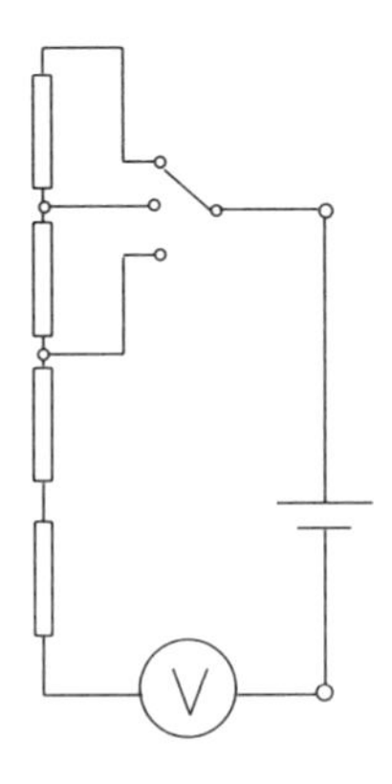

Figure 6.12

Example

A series chain of resistors is used to convert a meter movement, having a full scale deflection current of 0.5 mA and a resistance of 25 Ω, to read voltages on three ranges:

Range 1 0–10 V
Range 2 0–25 V
Range 3 0–50 V

$$I_m = 0.0005 \text{ A}$$
$$R_m = 25 \ \Omega$$

$$\therefore \quad U_m = 0.0125 \text{ V}$$

Range 1.

$$U_s = U - U_m$$
$$= 10 - 0.0125$$
$$= 9.9875 \text{ V}$$

$$R_s = \frac{U_s}{I_m}$$

$$= 19.975\ \text{k}\Omega$$

Range 2.

$$U_s = U - 10$$
$$= 15\ \text{V}$$

$$R_s = \frac{15}{0.0005}$$

$$= 30\ \text{k}\Omega$$

Range 3.

$$U_s = U - 25$$
$$= 25\ \text{V}$$

$$R_s = \frac{25}{0.0005}$$

$$= 50\ \text{k}\Omega$$

Try this

A meter movement has a resistance of 50 Ω and a full scale deflection current of 0.25 mA. Devise a method which incorporates a two-way switch whereby this instrument can be used to measure voltage on two ranges

(a) 0–250 V

(b) 0–500 V

Show a circuit diagram of the arrangement and the values of the resistors chosen.

Measurement of resistance

It is possible to use an ammeter and voltmeter to obtain the necessary values of current and voltage in order to calculate the resistance of a circuit or component.

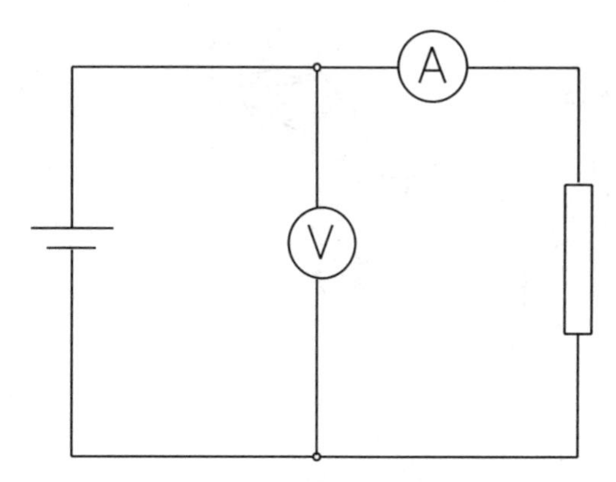

Figure 6.13 Voltage and current measurements for determining resistance

Direct measurement of resistance can be made using an ohmmeter.

The ohmmeter is a conventional meter movement with a power source, such as a battery, incorporated in the instrument.

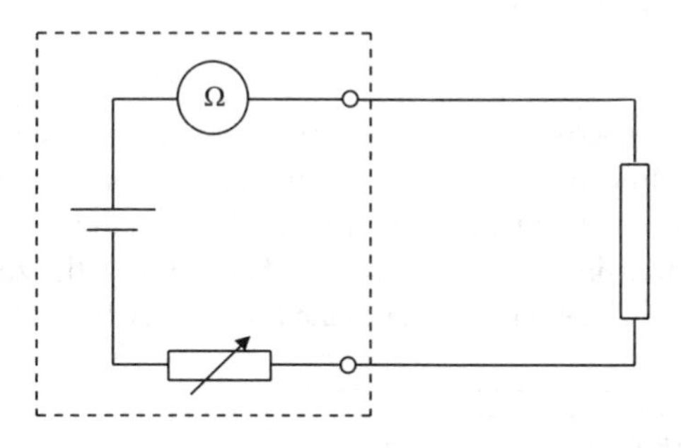

Figure 6.14 Determining resistance by measuring current

The current through the meter movement will be inversely proportional to the resistance of the circuit and the scale will be written showing the resistance on an inverted scale. This means that it will be non-linear and with the zero at the right hand end. Figure 6.15

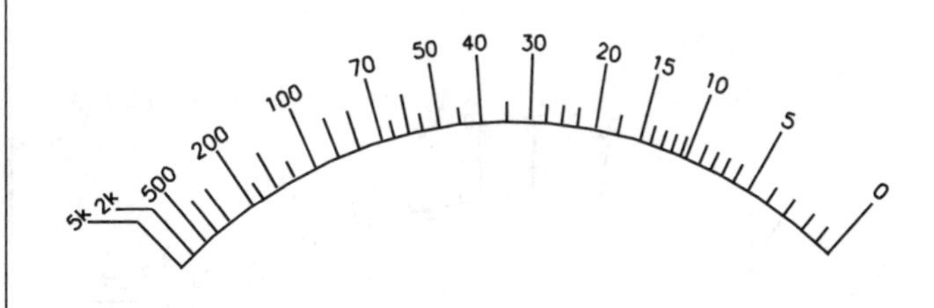

Figure 6.15

There has to be some form of adjustment on this type of instrument in order to compensate for internal resistance and battery condition. This is normally in the form of a variable resistor which can be adjusted for a zero reading with the leads short-circuited. Any resistance then connected between the leads will give the appropriate scale deflection.

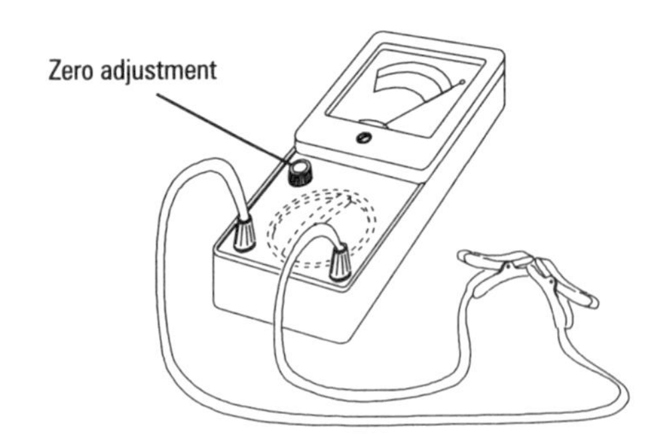

Figure 6.16

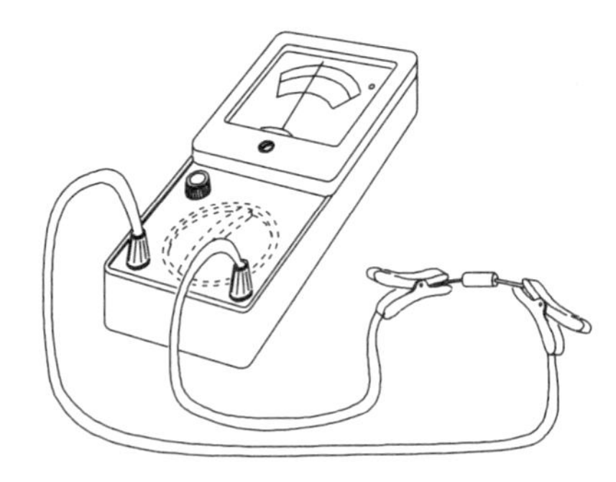

Figure 6.17

It is also possible to measure resistance by measuring the voltage drop produced by a known current in an unknown resistance as shown in Figure 6.18. In this case the meter scale will be non-linear but not inverted, that is to say, the zero will be at the left hand end of the scale (Figure 6.19).

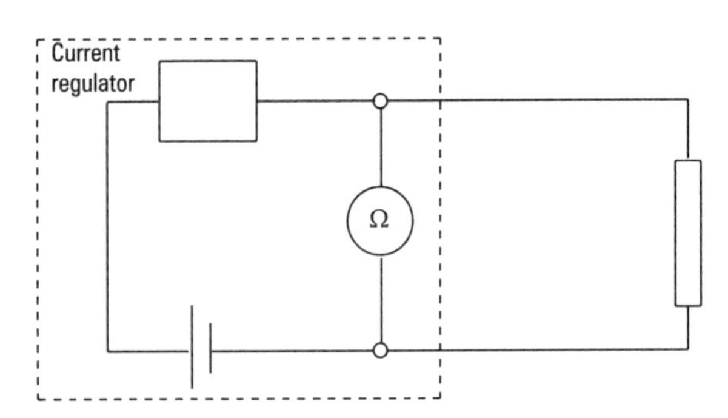

Figure 6.18

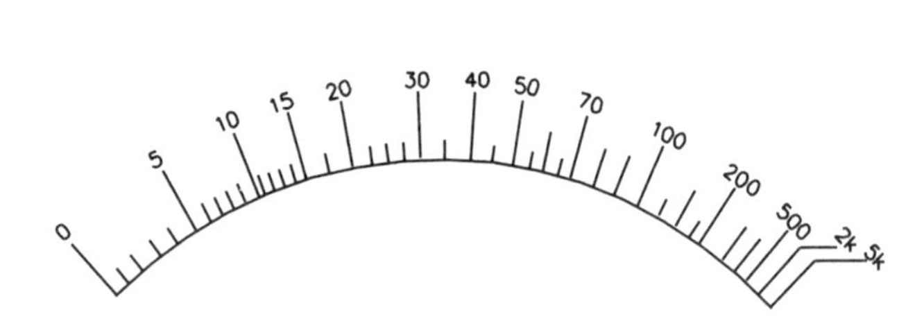

Figure 6.19

Bridge type resistance measuring instruments

The measurement bridge is dealt with in Chapter 3, "Circuit Theory", in this studybook. We will however now deal with the resistance bridge as a measuring device which can, if properly constructed, give very accurate results.

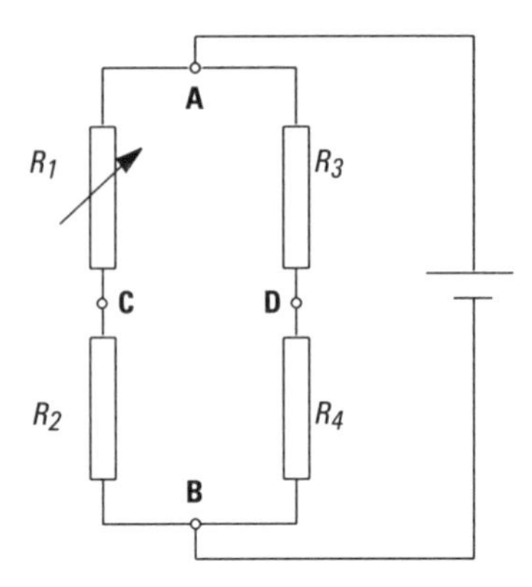

Figure 6.20

The circuit consists of two branches, each with two resistors in series. Current will flow through both branches and voltage drops will exist across all four resistors. Unless particular care has been taken with the choice of the resistor values we would expect to find a difference in voltage between the points **C** and **D**.

If however R_1 is adjustable then its voltage drop can be varied.

Then at some point the voltages at **C** and **D** will be made equal.

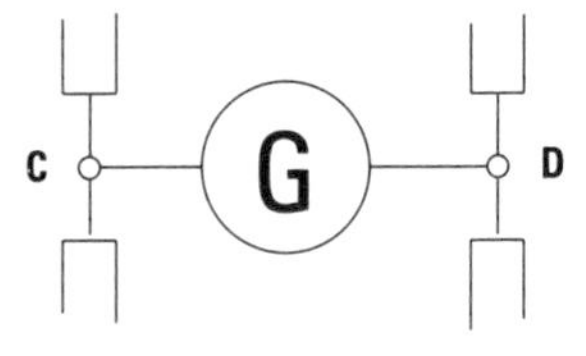

Figure 6.21

In order that this may be proved, we can put a very sensitive current reading instrument, known as a galvanometer, between these points and when there is no discernible current flowing we may then assume that the voltages are equal.

When the voltage of the upper half U_{A-C} equals U_{A-D} the lower half U_{C-B} will equal U_{D-B}.

At this point the bridge is said to be "balanced".

The condition for a balanced bridge is:

$$\frac{R_1}{R_2} = \frac{R_3}{R_4}$$

A resistance measuring bridge will have two variable resistors, R_3 and R_4, to make up the "ratio arm".

The other arm will consist of a third variable resistor R_2 and a pair of terminals for the unknown resistor R_x.

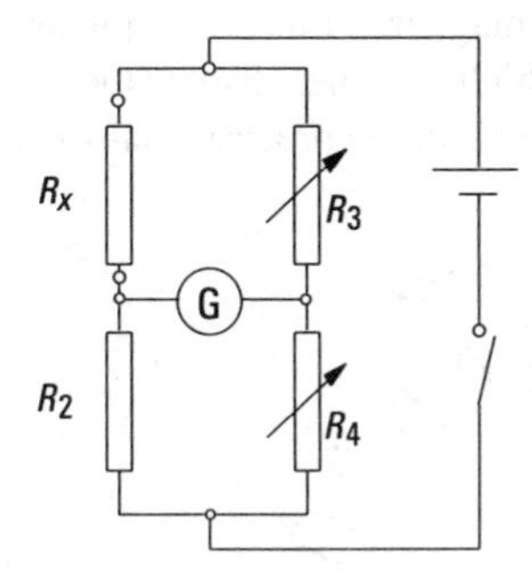

Figure 6.22

When the bridge is balanced, the galvanometer will indicate zero (the null condition) and the value of the unknown resistor can then be obtained by

$$R_x = \frac{R_2 \times R_3}{R_4}$$

Try this

When a bridge similar to Figure 6.22 is balanced the resistances are

$R_3 = 100\ \Omega$

$R_4 = 200\ \Omega$

$R_2 = 156.4\ \Omega$

Determine the value of the unknown resistor R_x

Measurement of temperature

A resistance bridge type instrument is frequently used to measure temperature in industrial processes.

For example, inside a drying chamber a detector bulb containing a fine platinum wire filament is located in the area where the temperature is to be measured. The resistance of the filament will vary with the temperature of the chamber. This resistance can be measured using the bridge but the ratio arm resistors will be high precision fixed value components and the settings of R_2 will then be directly calibrated in degrees to correspond to the temperature.

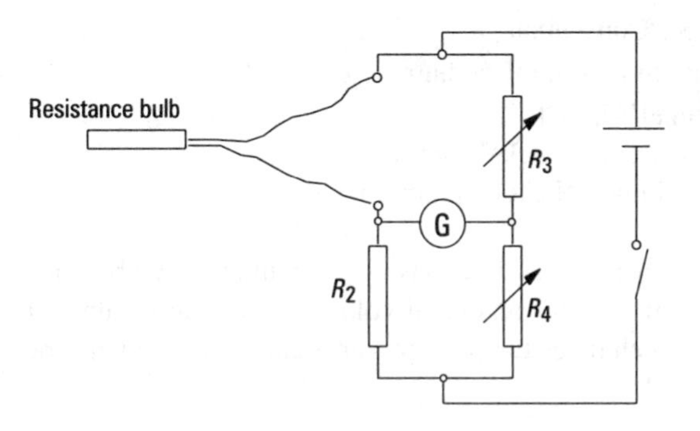

Figure 6.23

Temperature can also be measured by monitoring the current passing through a thermistor. The thermistor is a device the resistance of which is temperature sensitive. The current passing through the thermistor is measured on a milliammeter which has a modified scale calibrated in a temperature range which corresponds to the current in the monitoring circuit. Thermistors are available in positive and negative temperature coefficients. Those with a positive coefficient would cause a reduction in the deflection current with increasing temperature. If the thermistor has a negative temperature characteristic it would cause an increased current to flow corresponding to the increase in temperature. The indicating instrument would have to be calibrated according to the type of thermistor used and the scale would be normal or inverted as necessary.

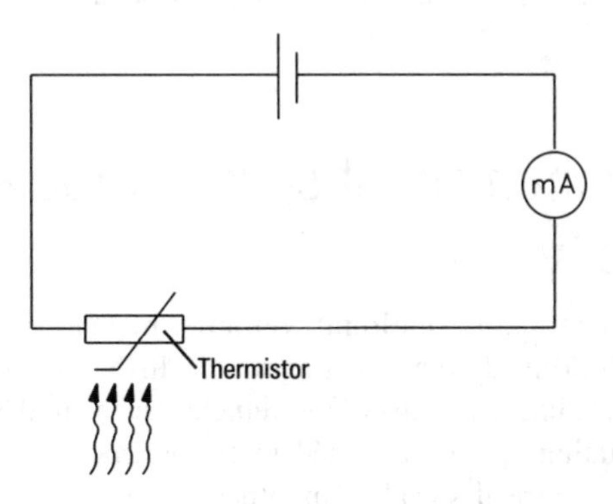

Figure 6.24

For very high temperatures, such as those encountered in furnaces and kilns, thermocouples are most commonly used. The thermocouple is essentially a fused junction between two metal wires which produces a small e.m.f. when heated. The e.m.f. is very small and can only be directly measured by means of a very sensitive instrument. The output voltage can, however, be amplified electronically in order to give a measurable output, or alternatively, it can be measured using a potentiometer which is a bridge type instrument capable of measuring the output of the thermocouple without drawing any current from it.

The most commonly used thermocouples use hot junctions which are

- Iron/Constantan
- Copper/Constantan
- Platinum/PlatinumRhodium
- Chromel/Alumel
  Chromel–NickelChrome,
  Alumel–NickelAluminium

There are many other junctions which could be used but they are either of too low an output voltage to be of any value, or they may melt in the temperatures encountered, which may be as high as 1500 °C.

An important application of the thermocouple is as a flame failure device associated with a gas or oil burner. The absence of an output from the thermocouple would indicate that the flame had gone out and the loss of this signal would cause a shut-down of the fuel supply. Further details of this can be seen in Chapter 7.

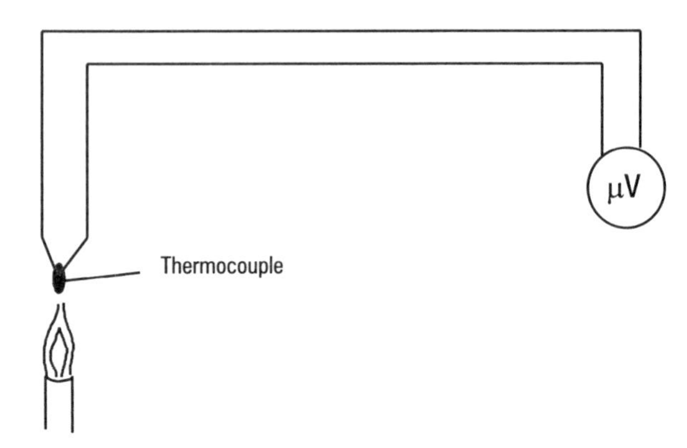

Figure 6.25 *Thermocouple used for temperature measurement*

Measurement of speed or radial velocity

Both linear and rotational velocity measurements are generally derived from a rotating source. In the case of a motor vehicle the linear velocity of the vehicle is measured by means of a rotating spindle attached to the gearbox, or from an induced electrical signal from either the engine or gearbox. The analogue types are simply an a.c. or d.c. generator attached to the rotating spindle.

The d.c. tachogenerator has a permanent magnetic field and its output is directly proportional to its speed. The a.c. tachogenerator is a small brushless alternator with a multi pole permanent magnet rotor. As with the d.c. type its output voltage is proportional to its speed. In most cases the output will be rectified and fed to a d.c. indicating instrument having a suitably calibrated scale.

An induction rotation indicator is a device frequently used in industry. This has a fixed pick up head which is positioned close to the teeth of a toothed wheel that has been mounted on the rotating shaft. As each tooth of the rotating wheel passes the tip of the pick up, a voltage is induced in the coil as a result of the change in magnetic flux. The pulses are fed to a receiving device which may then display the speed on a digital indicator or as a signal to the process controller.

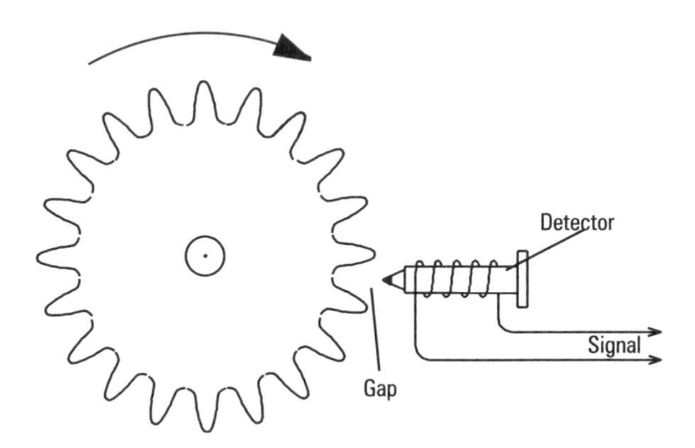

Figure 6.26 *Induction rotation indicator*

Measurement of power

In a simple d.c. circuit, the power can be measured by multiplying the readings obtained from an ammeter and a voltmeter.

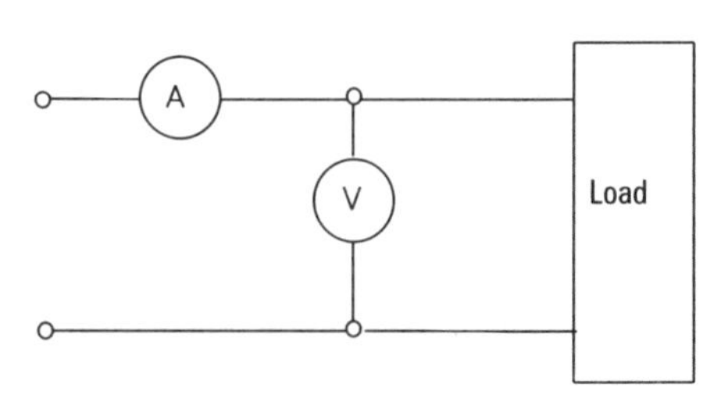

Figure 6.27

$$P = U \times I$$

In an a.c. circuit, this is no longer reliable since, as we have seen earlier in this studybook, the current and voltage may no longer be in phase with each other and the result obtained will be the VoltAmps rather than the active power of the circuit.

The electrodynamic (dynamometer) wattmeter connected as shown in Figure 6.28 will read the product of current and voltage at the same instant and will indicate the true power of the circuit.

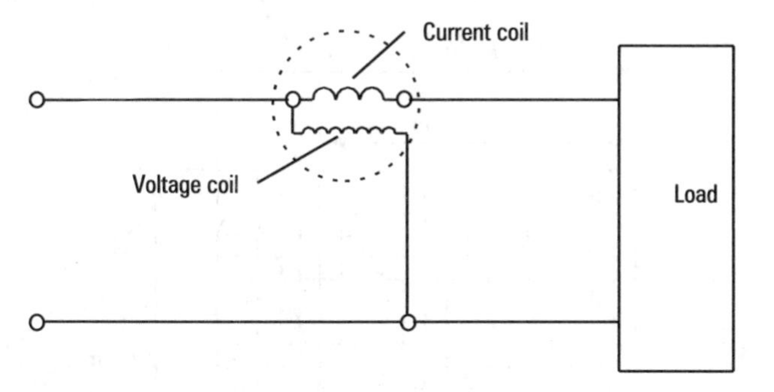

Figure 6.28 *Circuit diagram of wattmeter*

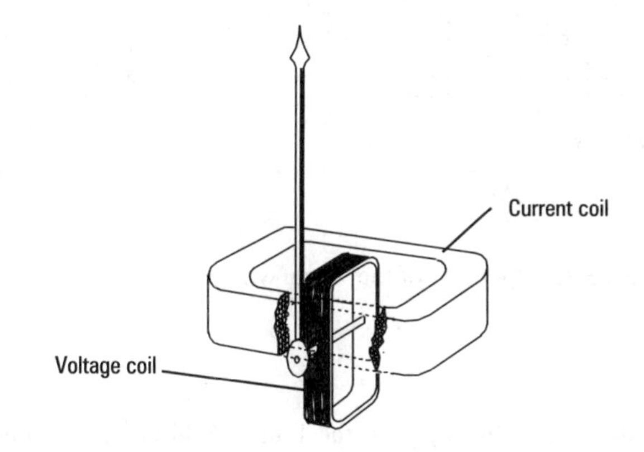

Figure 6.29 *Dynamometer*

As a portable instrument, the wattmeter has to be suitable for a wide variety of currents and voltages. If, for example, a wattmeter has three voltage and four current ranges this means that the deflection will have twelve possible interpretations.

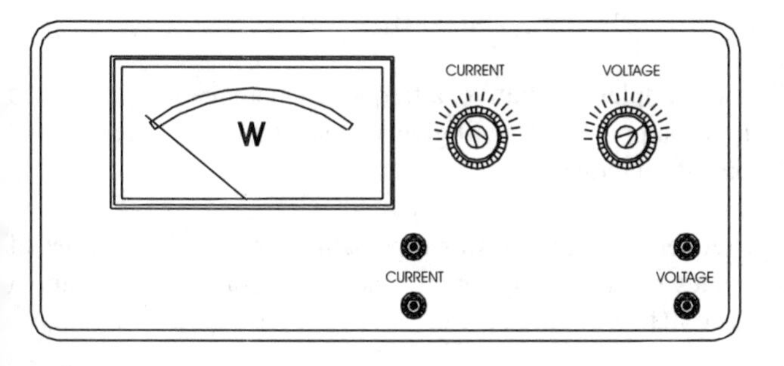

Figure 6.30 *A practical meter with voltage and current connections*

A scale of such complexity is obviously impractical so a single scale will be used and this will be read and multiplied by a factor determined by the current and voltage ranges selected.

Example

A wattmeter reads 0.7 on a scale of 0 to 1 when the current range is 0 - 1 A and the voltage range is 0 - 100 V. If the multiplying factor accompanying this choice of ranges is 100 then the power indicated is 0.7 × 100 = 70 Watts.

If however the instrument indicated 0.35 on the 0–1 A and 0–200 V ranges where the multiplying factor is 200 then the power indicated would be 0.35 × 200 = 70 Watts.

Try this

A wattmeter indicates 0.54 on a scale of 0 to 1 when the multiplying factor is 400.

(a) What is the power indicated by the instrument?

(b) If the current and voltage ranges are changed so that the factor becomes 250 what should the scale deflection read?

Power factor measurement

If separate ammeter and voltmeter readings can be multiplied to give the apparent power, or VoltAmps, and a wattmeter connected to the same circuit indicates true power then it is a simple matter to determine the power factor of the circuit.

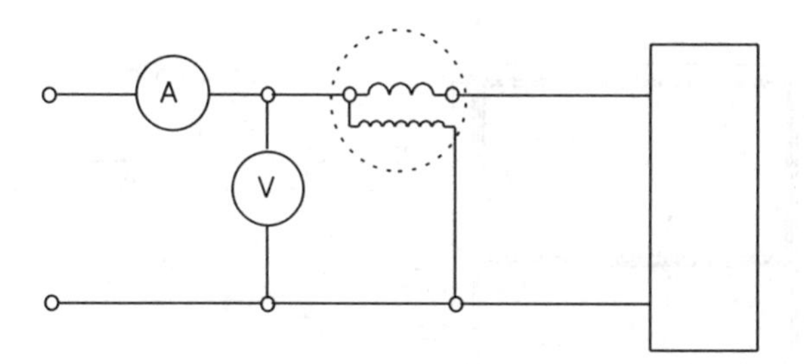

Figure 6.31

$$\cos \theta = \frac{P}{UI}$$

Example

From the circuit in Figure 6.31 the meter readings are

$$P = 750\text{ W}$$
$$U = 250\text{ V}$$
$$I = 5\text{ A}$$

$$\cos\theta = \frac{\text{power in Watts}}{\text{Volt Amps}} = \frac{750}{250 \times 5} = 0.6$$

This is a simple but cumbersome method of measuring power factor and it gives no indication whether the power factor is lagging or leading.

A purpose-made power factor meter looks at the current and voltage simultaneously and gives a reading on a scale which has a centre scale value of 1 (unity) and indicates the power factor lagging or leading on either side of this value

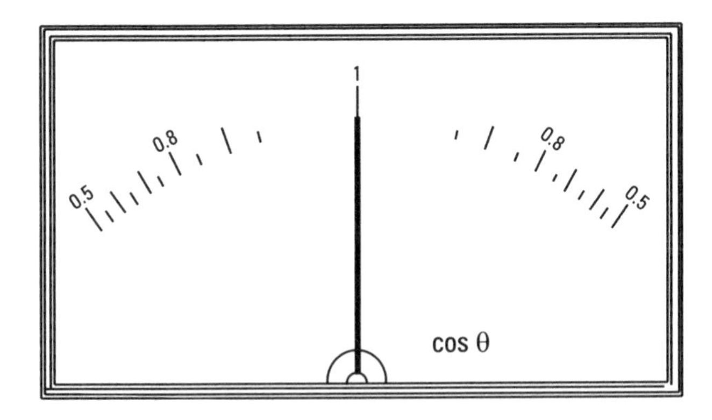

Figure 6.32

The voltage connection is usually made directly to the supply voltage but access to the current is normally by means of a current transformer.

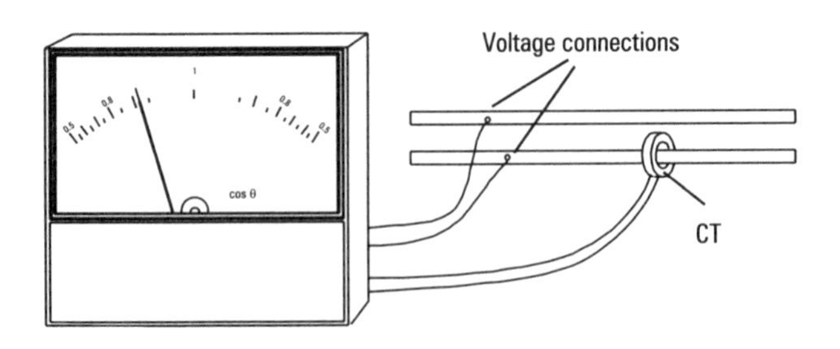

Figure 6.33

Details of current transformers are given further on in this chapter.

Measurement of frequency

One method of measuring frequency is to measure the length of one cycle of the alternating voltage as displayed on the screen of a cathode ray oscilloscope.

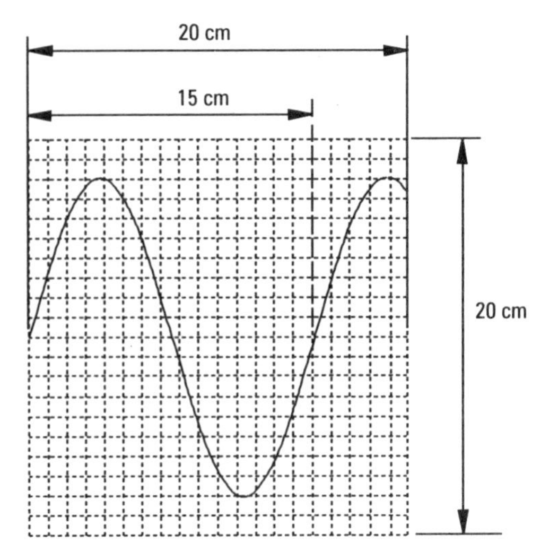

Figure 6.34

This gives the "period" of the wave, which is the reciprocal of the frequency.

Example

The sine wave displayed in the Figure 6.34 completes one cycle in a distance of 15 cm.

At a time base calibration of 1 ms/cm this tells us that one cycle has a period of 15 ms.

The frequency is therefore

$$f = \frac{1}{0.015} = 66.66\text{ Hz}$$

As a simpler alternative, a frequency meter can be connected to the supply which will give an instant and accurate read out on a digital display.

Frequency meters of this type actually count the number of cycles electronically and are accurate from 20 Hz to about 200 MHz with proper use.

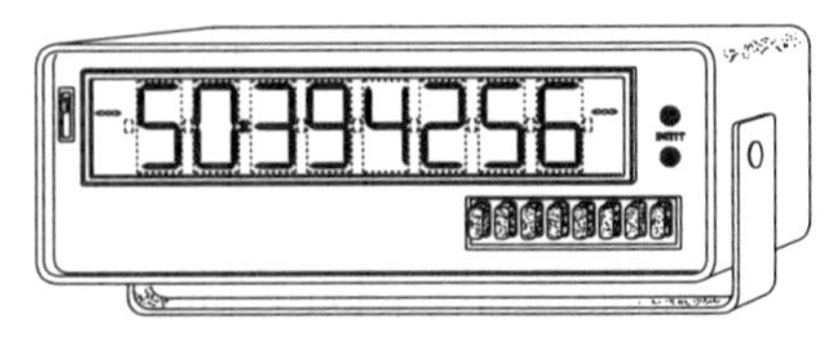

Figure 6.35

Try this

Calculate the frequency of the sine wave displayed below if the time base calibration is 0.5 ms/cm.

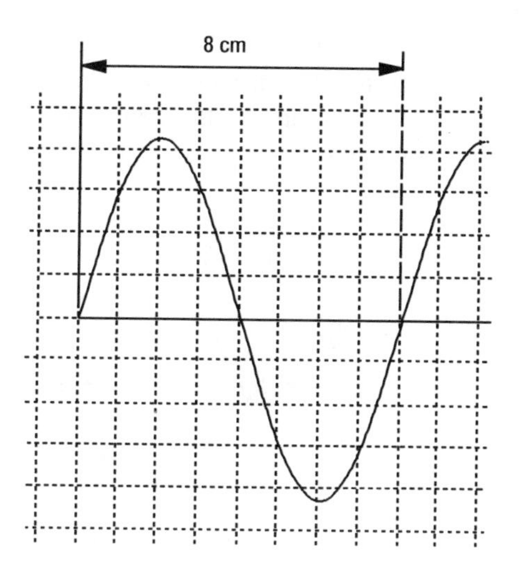

Digital instruments

Digital instruments fulfil the same function as analogue instruments in that they are used to take measurements and to convey the information to the user.

The operating principle is, however, quite different to the analogue types previously discussed.

The most obvious difference is the display. This is generally an alpha-numeric arrangement with seven elements although more sophisticated types are readily available.

Another difference is that the digital instrument requires a power source before it can be operated. This will be a battery or batteries, in the case of the portable instrument, or a power supply in the case of bench or panel instruments.

The operation of the digital instrument is the biggest fundamental difference. The measured signal will be a voltage, or in the case of current and resistance measurement, the voltage drop across a resistor of known value.

There are several methods used by different instrument manufacturers to determine the required values. One example is as follows.

The instrument uses a ramp generator to generate a voltage which rises at a uniform rate from zero to a maximum value before returning very quickly to zero (Figure 6.36).

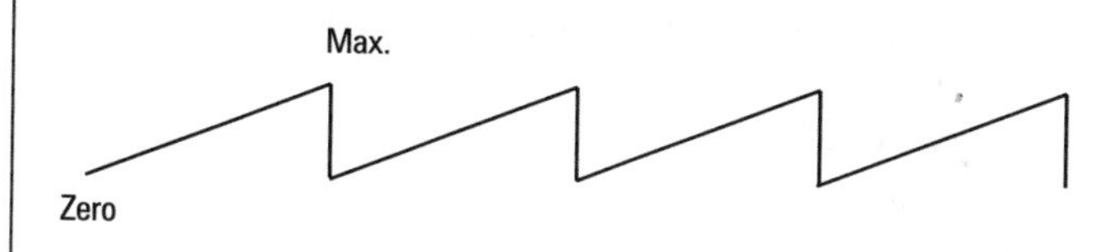

Figure 6.36 Output from the ramp generator

It also contains a pulse generator which generates pulses of a very precise frequency and duration continuously (Figure 6.37).

Figure 6.37 Output from the pulse generator

A circuit compares the input voltage to that on the ramp generator. From a point when the ramp generator starts at zero, a monitoring device counts the number of pulses from the pulse generator until the voltage on the ramp generator and the input voltage being measured match. At that point the counter stops.

4371

The result of the count is then translated into the required value and displayed in the appropriate form.

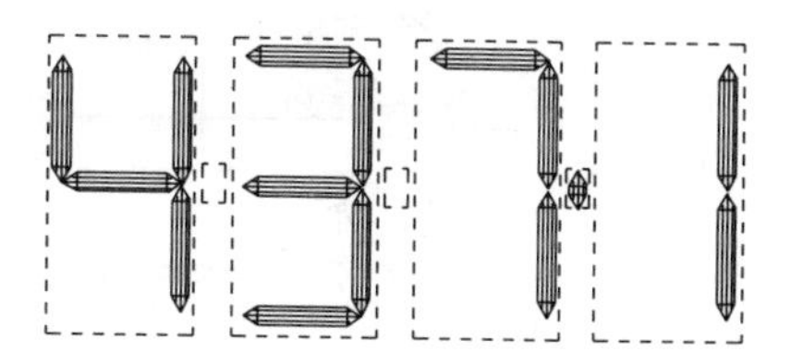

Figure 6.38 Digital display

Applications of measuring instruments

Instrument transformers

Current and voltage transformers are a common feature of fixed panel meter installations. The use of instrument transformers brings several advantages to the user, some of which are:

- Increased safety
- Practicality
- Economy

Increased safety

Instrument transformers isolate the instrument circuit from the main circuit. This is an absolute necessity for high voltage installations but even in low voltage installations this isolation can improve safety. By earthing the secondary of the instrument transformer this ensures that the instruments are as close as possible to earth potential.

Practicality

Where very high values of current or voltage are to be indicated, the use of instrument transformers allows meters to be installed at any convenient position and not necessarily at the point of application. Meter leads may be as long as is necessary and of small cross section with light insulation.

Economy

The secondary values are a matter of choice and for economy's sake these are normally low values of current and voltage. The instruments installed are of conventional construction, readily available and do not have any expensive special features.

Current transformers

Current transformers, or CT's as they are commonly called, consist of a doughnut shaped core around which is wound several turns of wire which make up the secondary winding. The primary winding is usually a single conductor such as a large cross-section cable or bus-bar.

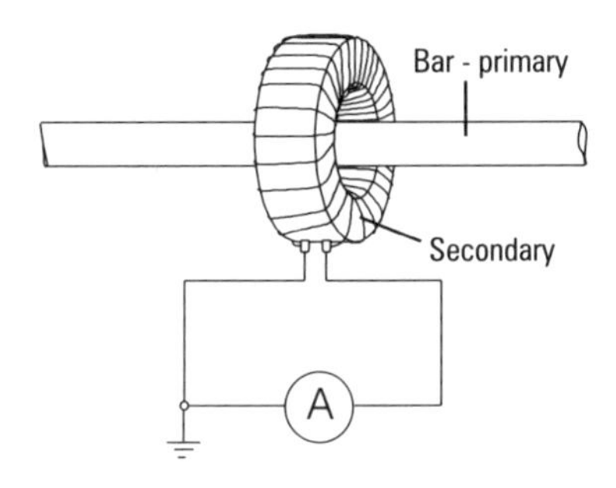

Figure 6.39

The secondary winding consists of a number of turns of insulated wire determined by the required output current.

For example if a current of 650 A is to be indicated on a scale of 0 to 1000 A using an ammeter whose full scale deflection is 5 A then the CT would need to have an effective ratio of 1000:5 (or 200:1).

The current transformer ratio may, in some cases, be modified by passing the primary through the transformer more than once, as shown in Figure 6.40.

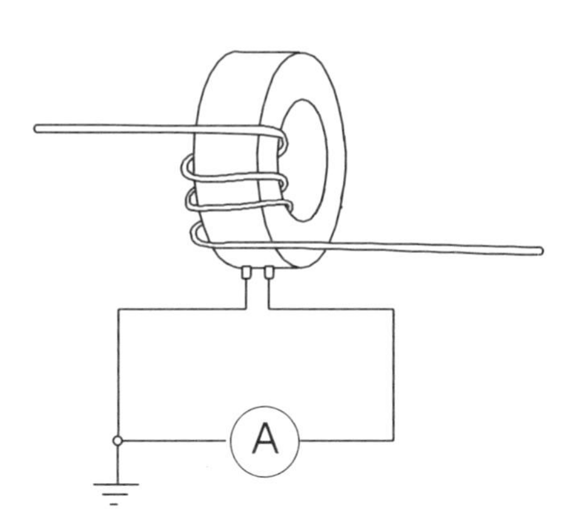

Figure 6.40

This is clearly not practical where very large primary conductors are being used but the practice may be adopted with smaller cables.

The loading of a CT is known as the "burden". For example, a CT secondary may deliver 5 A at a voltage of 3 V thus giving a burden of 15 VA at an impedance of 0.6 Ω including the resistance of the connecting leads.

The impedance of the secondary circuit must be taken into account and can be calculated from

$$Z = \frac{U}{I}$$

The manufacturer will be able to give an estimate of the correct burden for a CT, but these are available in a range of values between 5 and 50 VA per phase.

CAUTION

A CT must never be left in place on a loaded conductor without a secondary load of some description. A loaded primary conductor always produces a secondary voltage, and without a secondary load this can be high enough to cause an accident or even damage the CT.

If the ammeter has to be removed or disconnected for any reason always short-circuit the CT terminals first before disconnecting it. This will not harm the CT in any way and will prevent a dangerous situation from arising.

Hint
Wrap several turns of bare copper wire around the CT terminals before disconnecting the meter. These can be left on and cut away with your side cutters after re-connection.

If a switching arrangement is used to distribute the CT secondary to several ammeters situated in different locations, the switch must be of the make before break variety so that the CT secondary is not broken during the switching operation.

Voltage transformers

Voltage transformers, often abbreviated to VT's or less commonly PT's (potential transformers) are used to connect a voltmeter or voltage coil to a mains supply. The use of the VT is essential for high voltage measurement such as 11 kV or 33 kV, but even at 400 V or 230 V the isolation from a direct mains connection and a possible reduction to 110 V can lead to increased safety.

A voltage transformer is a straightforward double wound transformer of conventional construction and is not to be shorted out when not in use. The secondary terminals should however be insulated and enclosed to prevent accidental contact.

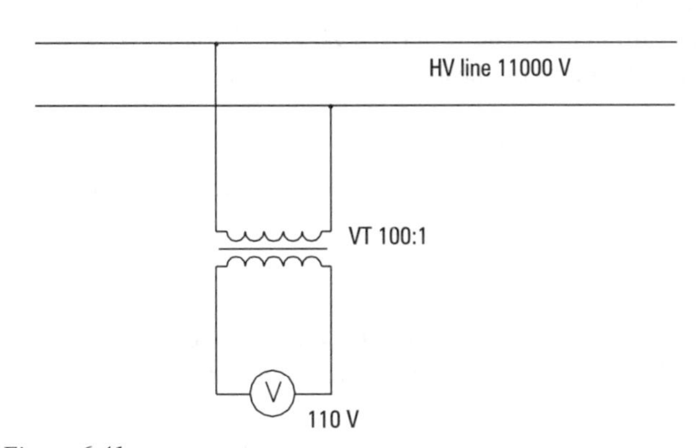

Figure 6.41

As in the case of the CT, a VT should be matched to its secondary circuit in terms of the appropriate "burden".

Burden
To calculate the burden of a measurement circuit, first note the current in the circuit, square it and multiply this by the impedance of the circuit. Or alternatively; measure the voltage across the transformer terminals and multiply this by the current in the circuit. In either case you will get a value in VoltAmps which should not be greater than the burden of the transformer.

Example
A CT has a secondary current of 15 A at an operating voltage of 2 V.

Burden = 15 × 2 = 30 VA

Try this
What is the burden of a CT which measures up to 25 A in a secondary circuit with an impedance of 0.06 Ω?

Moving coil instruments on a.c.

Moving coil instruments are primarily designed to measure d.c. values only. When used in conjunction with a rectifier to indicate a.c. values they respond to the AVERAGE and not to the r.m.s. value which is normally required.

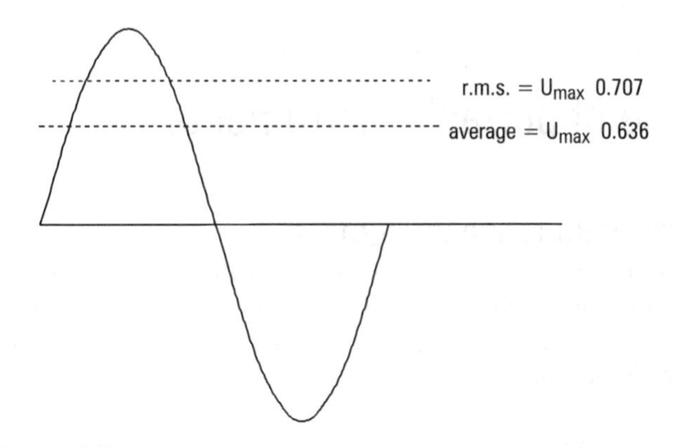

Figure 6.42

Since the r.m.s. value is greater than the average by a factor of 1.1 this adjustment must be made to the calibration of the instrument otherwise it will record a value which is consistently low by a corresponding amount. Moving iron instruments do not have this problem.

Example
An unadjusted moving coil instrument indicates a little over 54.5 A on an a.c. circuit. What is the true r.m.s. value of the circuit current?

r.m.s. current = average current × 1.1

= 54.5 × 1.1

= 59.95 (say 60 A)

Try this

What would be the reading on an unadjusted moving coil voltmeter connected to a 230 V r.m.s. supply?

NOTE

This is only true of pure sine waves, other waveforms present all sorts of problems in this area. This includes sine waves which are heavily affected by harmonics.

Use of portable instruments

The multi-range instrument

An electrician's kit is not complete without a multi-range instrument of some sort. It need not be an expensive, super quality, state-of-the-art piece of high technology, just a simple, reliable meter which is safe to use.

It should read voltage from a few volts up to say 500 V and resistance over several ranges so that the difference between short circuit and open circuit can be established.

Most multi-range instruments come with current ranges included but in practice, these are very seldom used.

Analogue instruments of this type are still popular but their delicate moving parts are not ideally suited to life in the back of a van.

The cost of digital multi-range instruments has come down considerably over recent years and some of very good quality have become quite affordable. There are also some "cheapies" around which should be viewed with some care.

The digital instrument is much more robust than its analogue counterpart due to the absence of delicately balanced moving parts. In addition to this, the digital instrument of this type is usually enclosed in a shock-proof plastic housing as compared to the glass and bakelite of the older analogue types.

Errors

The analogue instrument is still very useful in the workshop but care must be taken when measuring voltages on equipment which has high values of resistance incorporated in the circuit.

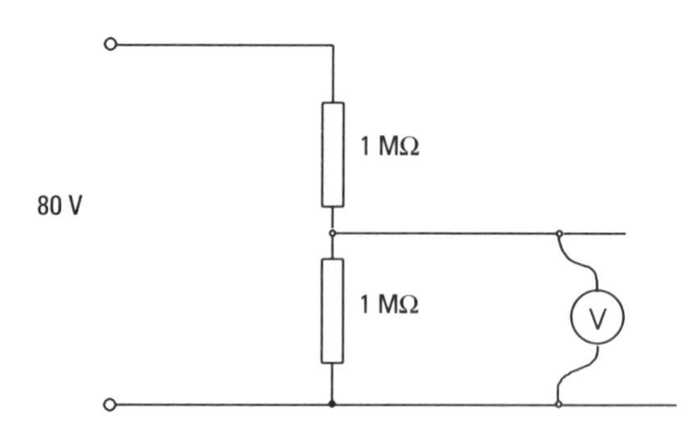

Figure 6.43

For example, in Figure 6.43, the true value of the output is 40 V but when the meter is connected this will fall to a lower value due to the current taken by the voltmeter.

To ensure greater accuracy of measurement, the voltmeter should have as high a resistance as possible. The instrument manufacturer will probably express this as an "ohms per volt ratio", which remains the same over all the ranges.

Example

An instrument is said to have a 20 kΩ / V ratio. What is its actual resistance on the 100 V range?

Meter resistance

$= \Omega/V \text{ ratio} \times \text{full scale deflection value}$

$= 20000 \times 100$

$= 2\ M\Omega$

Yes, this is a high value but when it is connected across another 2 MΩ resistor the circuit resistance is halved.

A digital instrument has its own power source and measures the voltage by comparing it with another. For this reason, digital instruments tend to be more reliable in such circumstances.

Try this

A multi-range instrument has a meter movement which works on 2000 Ω/V. Calculate

(a) the resistance when the meter is on the 25 V range
(b) the current flowing in the meter when measuring 15 V on the 0–25 V range.

NOTE ON ISOLATION PROCEDURE

It has been strongly recommended by Her Majesty's Health and Safety Inspectorate that voltage indicators should be used in preference to multi-range instruments when testing for the presence of voltage.

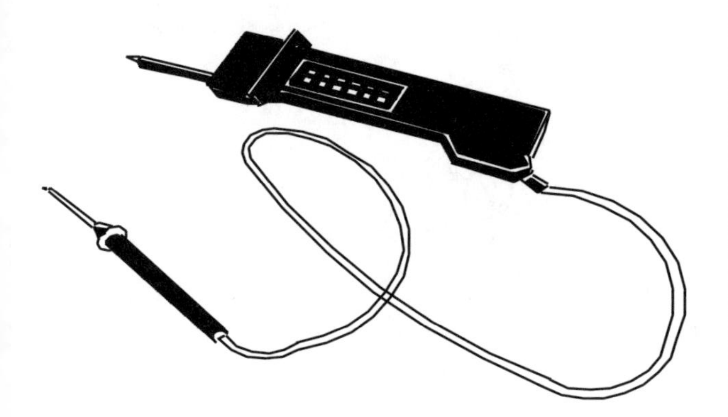

Figure 6.44 Voltage indicator

Before using a voltage indicator to verify that a circuit has been isolated it is essential that the operator follows the procedure as laid down by the JIB for the Electrical Contracting Industry. You will learn from these instructions that the instrument must be checked against an independent source voltage immediately before and after the testing stage. This is to verify that the instrument was working at the time of the test.

Try this

Draw a flowchart showing an acceptable procedure for isolation.

On-site current measurement

It is virtually unheard of to take direct measurements of current from on-site, power-using equipment. The whole process is fraught with danger and difficulty, and in most cases, not worth considering.

A "clip-on" ammeter is a device which can do the job most effectively by measuring the strength of the magnetic field around a single conductor in exactly the same way as a CT measures the current in a bar primary.

A lever opens the jaws of the instrument and allows it to be placed around the conductor whose current is to be measured.

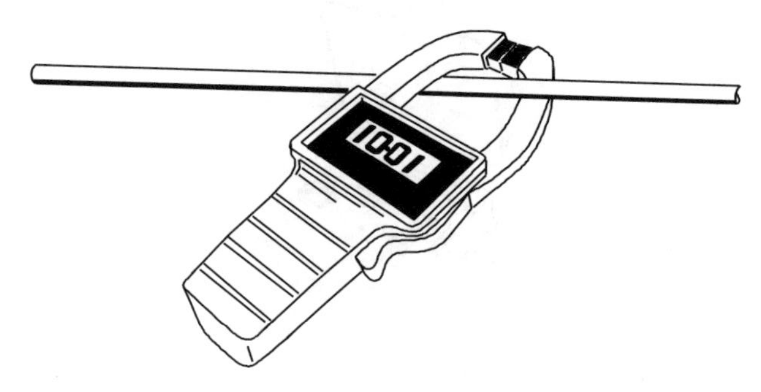

Figure 6.45

These instruments usually have a selection of ranges up to several hundred Amperes and are very simple to use.

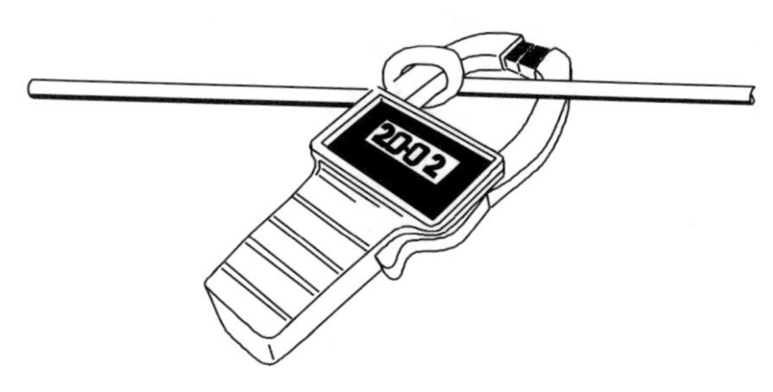

Figure 6.46

For the measurement of small currents it may be necessary to coil the conductor around the tongs several times in order to get a sensible scale deflection. The reading must then be divided by the number of turns.

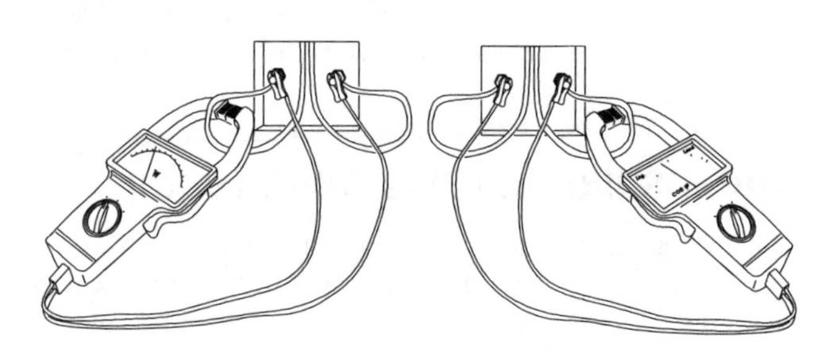

Figure 6.47 Wattmeter and power factor meter with clip-on current tongs

Continuity testers

Simple continuity tests can be made by using the ohms range of a multi-range instrument. This will give a reliable indication of conductor continuity (or the lack of it) and whether or not a fuse has "blown". Measurement of resistor values can also be carried out, although at the far end of a non-linear scale this can be less than ideal.

With resistance measurement, care must be taken to ensure that there is no voltage in the circuit which is coming from another source, as this will undoubtedly affect the reading and may damage the instrument.

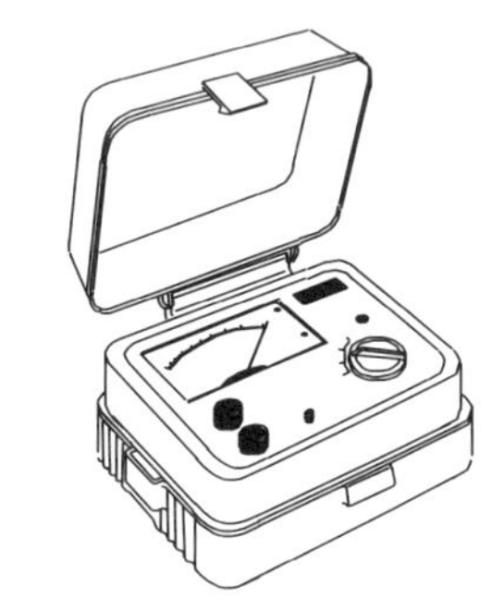

Figure 6.48 *Continuity tester*

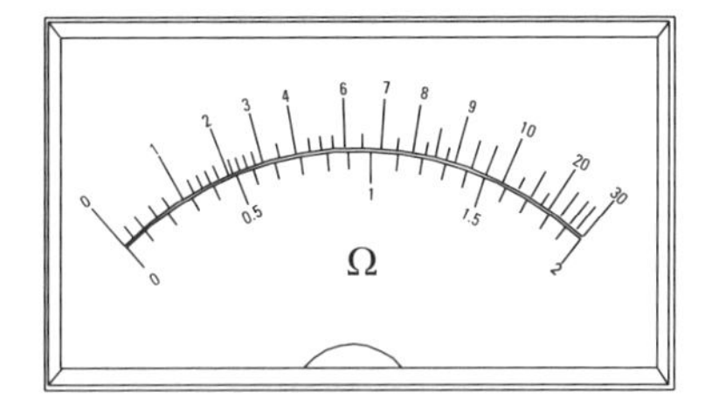

Figure 6.49

The purpose-built continuity tester is designed to read the low resistance values such as are typical in supply and earthing conductors.

For installation testing purposes, levels of resistance have to be reliably measured at values of less than 0.1 of an ohm.

This is very demanding on the instrument and the operator, and requires that the utmost attention is paid to such details as the resistance of test leads and the effectiveness of connections.

The insulation resistance tester

The insulation resistance tester is an instrument which is specifically designed for that purpose. It must not be used for continuity testing as its range is far too high and a value of several hundred ohms can produce a "zero" reading.

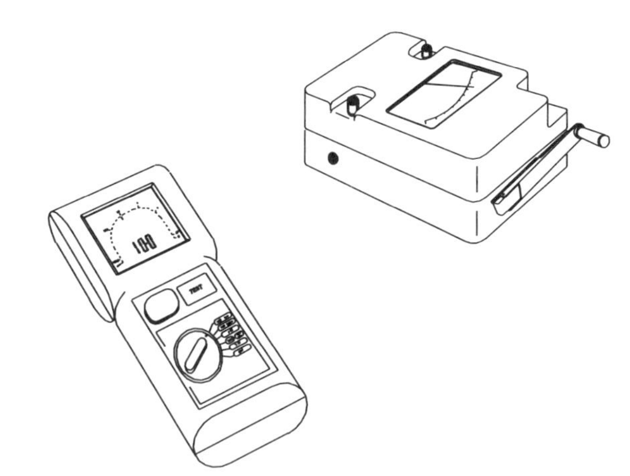

Figure 6.50 *Insulation resistance testers*

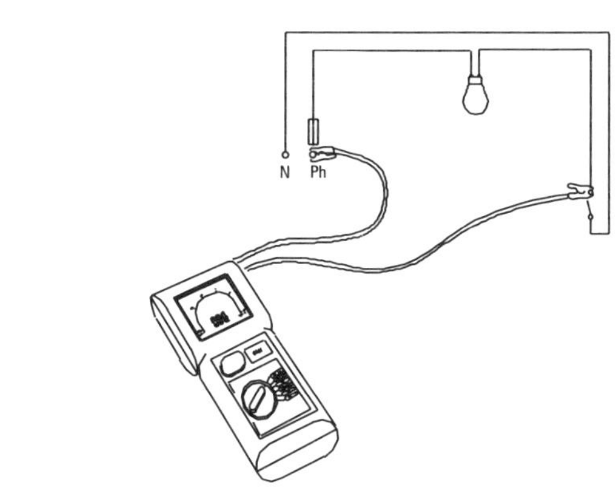

Figure 6.51

A common falsehood which arises is when testing the polarity of a switched circuit using this instrument on the megohms range. The zero reading might convince the operator that the polarity is correct in a circuit which is clearly incorrect due to the instrument being unable to measure the resistance of the lamp filament.

In a dual-scale tester it is quite correct and proper to select and use the continuity range for continuity tests.

Insulation resistance tests are carried out at voltages of 250 V, 500 V and 1000 V as required by BS 7671:1992 Requirements for Electrical Installations, otherwise known as The IEE Wiring Regulations.

Because the insulation test voltage is likely to be considerably higher than the normal operating voltage it is important to ensure that circuit components which are in danger of being damaged, are disconnected for the duration of the tests.

Before conducting an insulation resistance test it is vital that the mains supply is first disconnected and made safe by a proper isolation procedure.

Earth fault loop impedance testing

The earth fault loop impedance tester is an instrument which connects a 20 Ω resistor between the phase conductor and the earth connection at the test point. The current flowing in this test circuit (10 A to 20 A) is converted by the instrument's internal circuitry and displayed on the scale as a reading in ohms. The value shown is the earth fault loop impedance at that point in the circuit under test conditions.

That is to say, it measures the impedance of the whole earth fault loop, originating at the nearest power source, usually a sub-station transformer, through the phase conductors right up to the test point and returning to the same source by the earth return path.

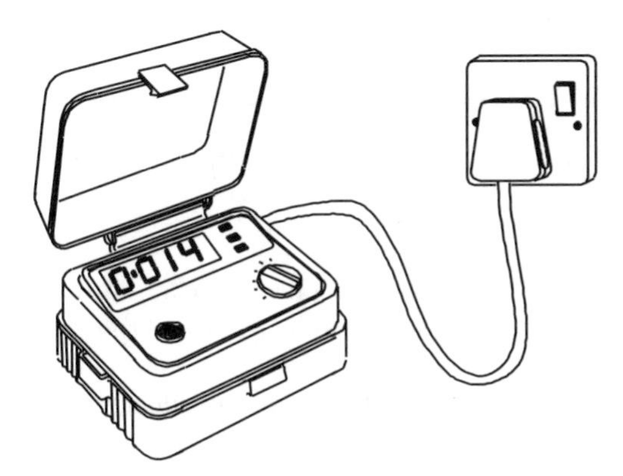

Figure 6.53 *Earth fault loop impedance test on a socket outlet*

It is not good practice to measure earth fault loop impedance in a circuit which is protected by an r.c.d. and for testing purposes this device can be bridged by means of temporary links for the duration of the test.

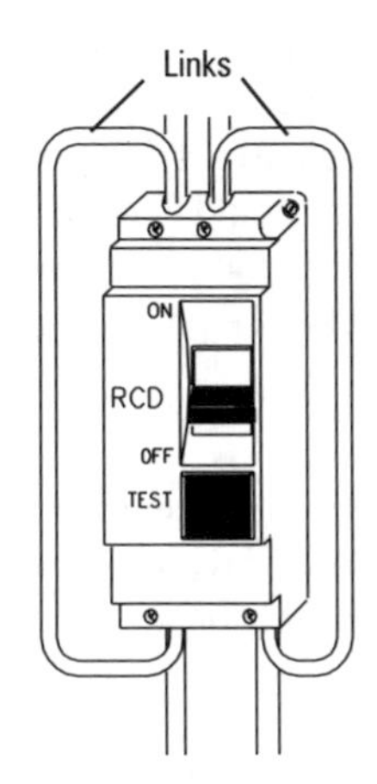

Figure 6.54

There are however loop impedance testers which can, with the aid of microprocessor circuitry take a reading of earth fault loop impedance without tripping any conventional passive device. These have an extended range of up to 2000 Ω which can actually measure the loop impedance through the earth electrode of a TT system and this is a very useful feature.

Most testers are designed to be inserted in a socket outlet for the convenient testing of ring and radial circuits. For other circuits, particularly fixed equipment, it is essential that the tester is connected to the phase and earth connections of that circuit. It is not appropriate to use a convenient socket outlet to supply the instrument and use the remote earth probe to the exposed metalwork of the equipment under test if this is fed from a different circuit.

The loading on the test resistor is high and for this reason it is unwise to make too many tests within a short period without allowing the instrument to cool down between tests.

The instrument normally incorporates a device which will indicate that

- the circuit polarity is correct, and
- an earth connection is present.

Do not proceed with the test unless the instrument indicates that it is safe to do so.

Earth electrode testing

The testing of earth electrodes on site can be carried out using an instrument which passes an alternating current through the soil and uses a probe to detect the fall in voltage between the electrode under test and a second current electrode. The power source is independent of the mains supply and is generated from an internal battery as shown in Figure 6.55, or by means of a hand-cranked generator.

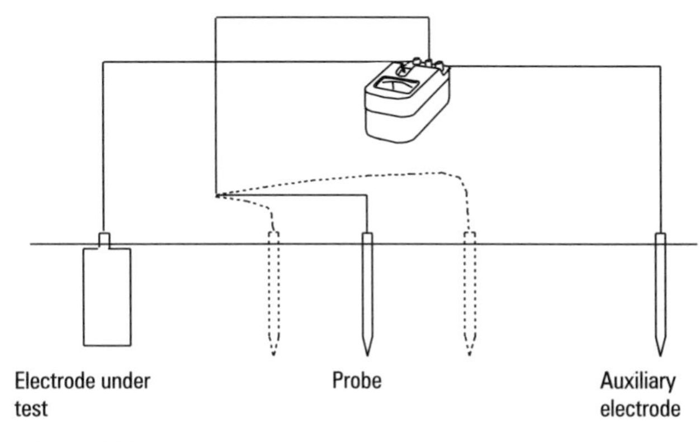

Figure 6.55

Instruments of this type will give a direct read-out in ohms which will be equivalent to the resistance between the electrode under test and the surrounding soil.

An enterprising electrician can construct his own test set from a 12 V bell transformer, an ammeter, a voltmeter and a couple of probes, as shown in the circuit in Figure 6.56.

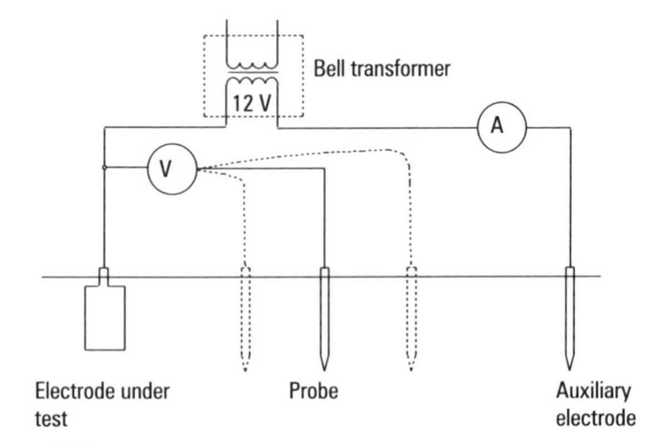

Figure 6.56

If a stable voltage has been obtained at several points approximately mid-way along the line between the electrodes, this value may then be divided by the current reading of the ammeter and the result will be a reliable indication of the electrode resistance.

Try this

In the circuit in Figure 6.56, the current indicated on the ammeter is 55 mA and the probe indicates voltages with an average value of 4.125 V.

What is the resistance of the earth electrode?

The R.C.D. tester

This instrument will measure and display the time taken for an r.c.d. to trip at a selection of pre-determined test currents.

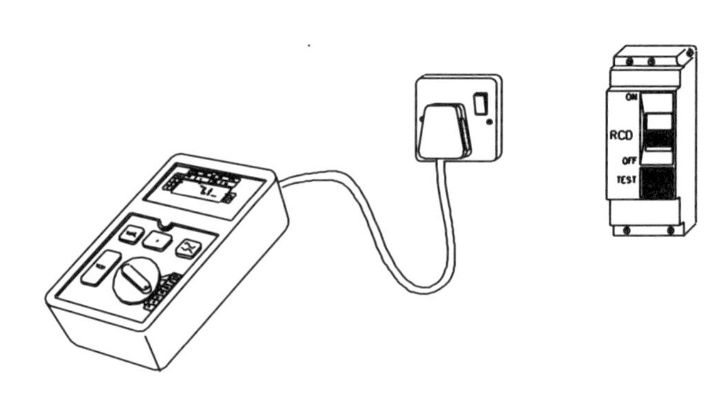

Figure 6.57 Testing an r.c.d. from a socket outlet

Method

1. Use the mains lead supplied with the instrument and insert this into a socket outlet fed from a source protected by the r.c.d. under test.
2. Check that the indicators on the instrument show that it is safe to proceed.
3. Set the scale to – trip current × 0.5.
   For a 30 mA trip this will be 15 mA. Press the test button and the r.c.d. should not trip. The display will show the "test time expired" symbol .
4. Reverse the phase angle from 0° to 180° and repeat. The trip should not operate.
5. Set the scale to – Trip current × 1.
   Press the test button. The trip should operate and the time elapsed will show on the display. This should be no more than the maker's recommended interval. Reverse the phase angle and repeat.
6. Set the current scale to – Trip current × 5. Press the test button and the trip should operate. The time elapsed will show on the display and in this case it must be no more than 40 ms. Reverse the phase angle and repeat.

Note

It is a requirement of BS 7671: Regulation 412-06-02 (ii) that a 30 mA r.c.d. should operate within 40 ms at 150 mA if it is to give any measure of protection against direct contact.

Measurement of three phase power

The power consumption of three phase unbalanced loads is normally treated as the sum of the power in each of the phases measured separately.

Three single phase wattmeters connected as shown, Figure 6.58, either directly or by means of CT's and VT's will give the indications required.

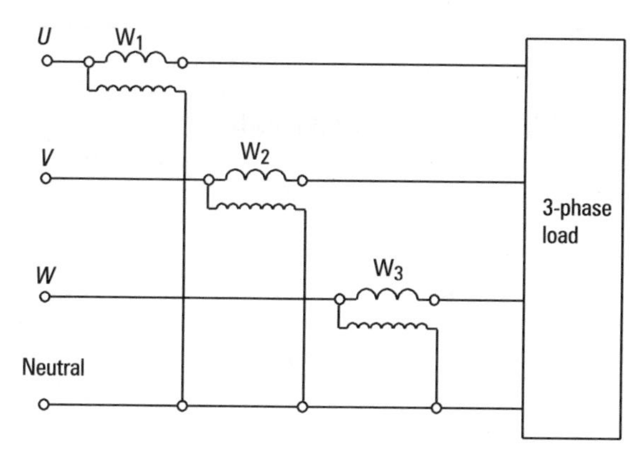

Figure 6.58

Two wattmeter method

For a balanced three phase, three wire load the two wattmeter method will indicate the total power in the load.

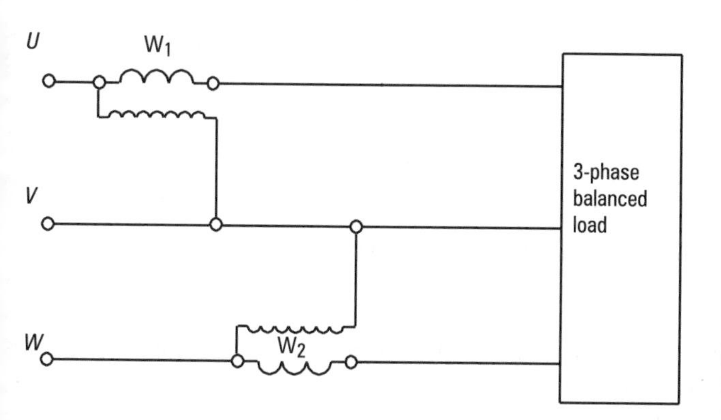

Figure 6.59

$$\text{Total Power} = W_1 + W_2$$

At unity power factor, the wattmeter readings will be identical, i.e. each will indicate exactly half the power in the load.

Indication of power factor by the two wattmeter method

At power factors other than unity it will be seen that the reading on one wattmeter has increased as the other has decreased so that the total power is still $W_1 + W_2$ but there is clearly a difference between them.

Due to the fact that both wattmeters use the third phase as a reference voltage, any phase shift in the current causes it to move closer to the voltage in one line and further away in the other.

It can be proved that the tangent of the phase angle can be found by

$$\tan\phi = \frac{\sqrt{3}\times(W_2 - W_1)}{W_2 + W_1}$$

i.e. "$\sqrt{3}$ times the difference over the sum = $\tan\phi$"

Having got the tangent by this method, find its angle and then its cosine and you will have the power factor.

Example

A three phase balanced load is connected for power measurement using the two wattmeter method.

W_1 reads 500 W
W_2 reads 1000 W

Determine
(a) the total power
(b) the power factor

(a) Total power $= W_1 + W_2 = 1500$ W

(b) $$\tan\phi = \frac{\sqrt{3}\times(1000-500)}{1000+500}$$

$$= 0.577$$

$$\phi = \text{inv. tan } 0.577$$

$$= 30°$$

$$\cos\phi = \cos 30$$

$$= 0.866$$

TIPS

There are several landmarks in this method which are as follows:

If $W_2 = W_1$ p.f. is unity

If $W_2 = 2W_1$ p.f. is 0.866

If $W_1 = 0$ p.f. is 0.5

If W_1 reads backwards p.f. is less than 0.5

i.e. it is necessary to reverse the current or voltage coil connections to get a reading.

Try this

Using the two wattmeter method calculate the total power and power factor of the load, given that W_1 reads 400 W and W_2, 1200 W.

Exercises

1. Measuring instruments for high voltage and high current circuits are usually fed via instrument transformers.
   (a) Give three reasons for using this system.
   (b) Draw circuit diagrams to show how
      i) a wattmeter may be connected to measure the power in a single-phase high voltage circuit. The circuit should show all necessary protection and safety connections.
      ii) two single-phase wattmeters may be used to measure the power in a three-phase, three wire system.
   (c) Two single-phase wattmeters connected to measure the power input to a three-phase circuit read 350 W and 2250 W respectively. Determine the power supplied and the power factor when
      i) BOTH readings are positive
      ii) ONE meter reads backwards

2. (a) Two voltmeters A and B give identical readings when connected to a 240 V supply. When the same meters are used to check the output of a high impedance stabilised voltage unit, voltmeter A indicates a voltage of 200 V and voltmeter B indicates 160 V . Explain a probable cause for this difference in the readings.
(b) When measuring the output of a full-wave bridge rectifier, a moving iron instrument indicates a current of 7 A flowing. When a moving-coil instrument is used for the same measurement it indicates 6.3 A. If both instruments have been recently calibrated explain the reason for the difference in the readings.

3. A wheatstone bridge circuit consists of four resistors PQ, QR, RS and SP connected to form a closed loop. A galvanometer is connected across P–R and a cell across Q–S. The cell has an e.m.f. of 1.5 V. When the bridge is at balance PQ = 9 Ω, QR = 3 Ω and RS = 2 Ω. Draw a circuit diagram and calculate the
(a) value of SP
(b) current in PQ
(c) current in QR

4. (a) For each of the following, state a suitable type of measuring instrument:
   i) d.c. and average quantities
   ii) d.c. or a.c. and r.m.s. quantities
   iii) voltage across a capacitor
   iv) simultaneous measurement or two related quantities
   v) d.c. or a.c. quantities where surges may give false readings
   vi) instantaneous and peak values and frequency

   (b) Compare the scale of a moving-coil instrument with that of a moving-iron instrument and comment on the relative accuracy obtained when reading values from these scales.

   (c) Explain, with the aid of a circuit diagram, how a 5 A moving-iron instrument is used to measure an a.c. current of 400 A.

7

Electronics

Complete the following to remind yourself of some important facts on this subject that you should remember from the previous chapter.

1. State the three principal advantages associated with the use of instrument transformers.

2. What is the numerical value of the factor used to enable moving coil instruments to be used on a.c. circuit measurements?

3. In a 2 wattmeter test W_1 reads 560 W and W_2 912 W. What is the power factor?

On completion of this chapter you should be able to:

- identify waveforms on an oscilloscope
- list applications for electronic components
- identify given components from their packaging
- describe the output from given circuits
- determine the results from the use of logic gates
- describe the procedures for basic fault finding

To quote from the City & Guilds 2360 Course C syllabus:

"The aim of this section is to give a basic understanding of electronic applications in electrical installation work. It is intended that much of the content will be taught within a workshop/laboratory environment."

The whole area of study which can be covered under the heading of "Electronics" is too vast to be covered in one short chapter.

The purpose of this chapter is, therefore, to explore certain topics which come within this sphere in such a way as to complement the student's workshop/laboratory activities.

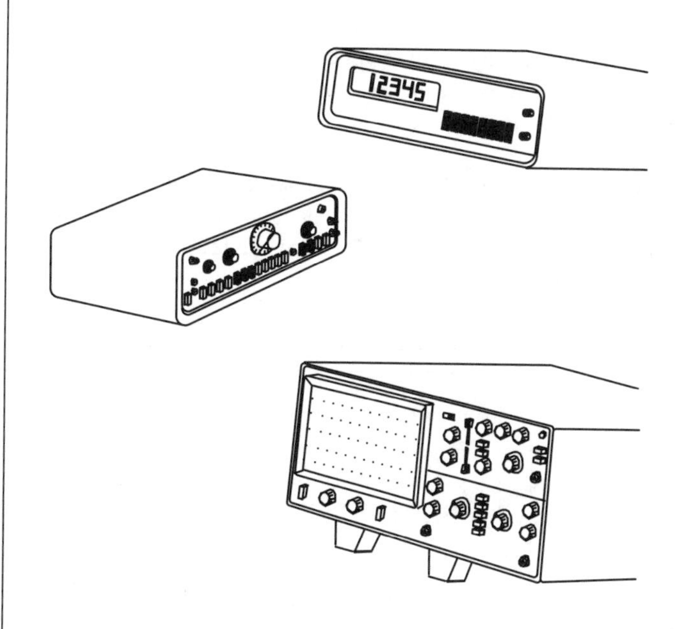

Figure 7.1

The cathode-ray oscilloscope

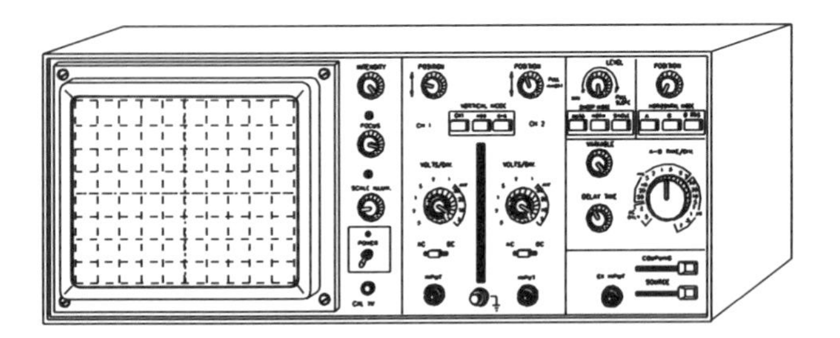

Figure 7.2

The business end of the Cathode Ray Oscilloscope (CRO) is the cathode ray tube. This performs the same function as it does in a T.V. set, insomuch as it displays a picture. The screen is divided up into one centimetre squares and this pattern is known as the "graticule". The graticule is centred on the intersection of the two axes.

The X axis is horizontal and is normally used as a time scale. The time scale is variable and the calibration can be selected using the time base selector switch which has a range of settings from seconds per centimetre, through milliseconds per centimetre, to microseconds per centimetre.

The trace on the CRO screen is actually the path taken by a spot of light on the screen. With no signal voltage connected and the time base calibration on its lowest setting you will see the spot coming on to the screen on the left and disappearing out at the right-hand edge only to appear again on the left. As you speed up the time scale the dots move faster until they merge to form a continuous line.

The Y axis is vertical and this also has a calibrated setting in Volts per centimetre on the Y, or amplitude, control. The sensitivity can be adjusted right down to a few microvolts per centimetre.

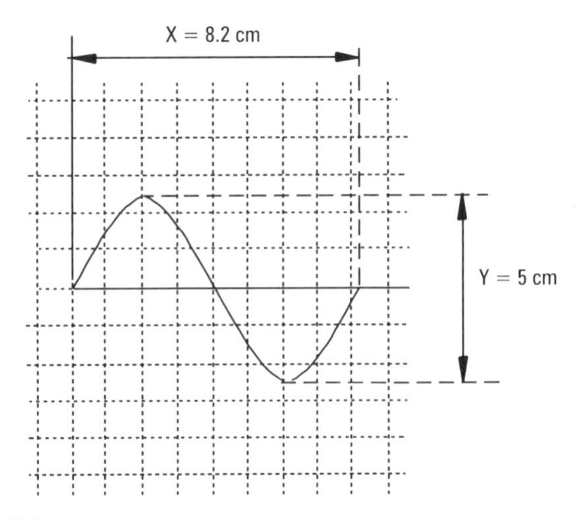

Figure 7.3

The trace on the screen, Figure 7.3, shows a sine wave with the graticule superimposed on it. From this, and the settings indicated on the controls, we will try to find out all we can about the sine wave displayed. From the trace, we see that the length of the sine wave is 8.2 cm and the amplitude or peak-to-peak height is 5 cm.

Time base

The setting on the X control is 2 milliseconds per centimetre therefore the time taken to complete one complete cycle is 8.2 cm at 2 ms per centimetre

∴ the period of the wave is

$$0.002 \times 8.2 = 0.0164 \text{ seconds}$$

The frequency is the reciprocal of the period

$$f = \frac{1}{0.0164}$$

$$f = 60.97 \text{ Hz}$$

Note

This type of measurement does not warrant such accuracy so the result would probably be taken as 61 Hz.

Amplitude

The overall peak-to-peak voltage is 5 cm at 1 Volt/cm
= 5 V

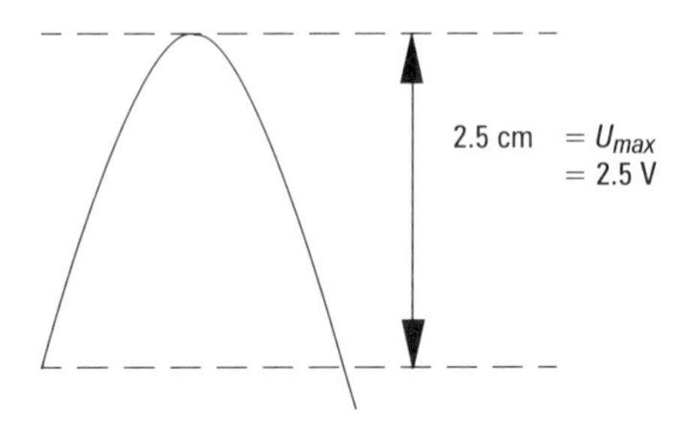

Figure 7.4

∴ the peak value on a half cycle is 2.5 V.

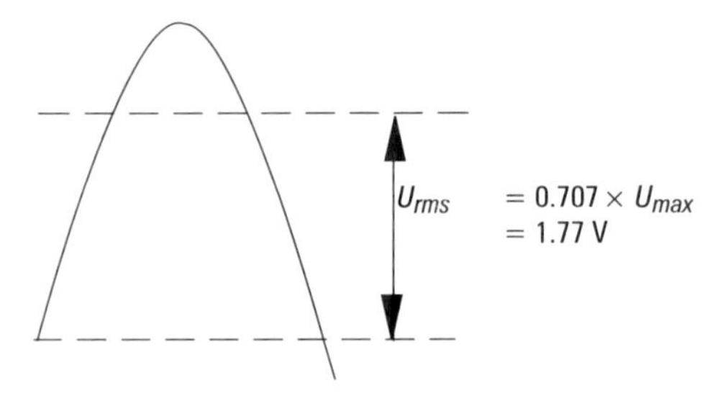

Figure 7.5

and the r.m.s. will be 0.707 of this value

i.e. U (r.m.s.) = 1.77 V

Try this

Determine, from the CRO trace,

(a) the r.m.s. voltage

(b) the frequency

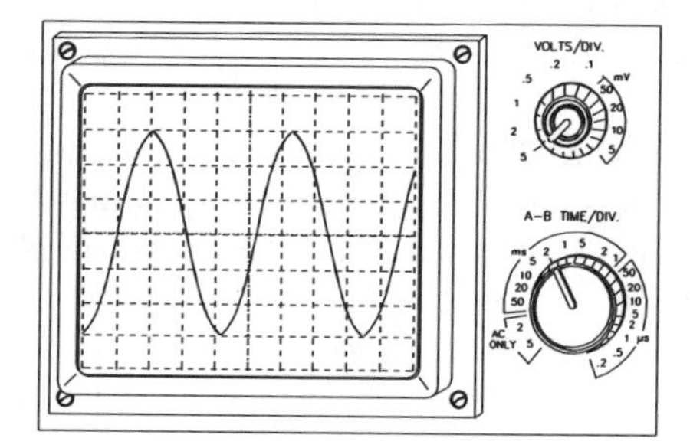

Y = 5 V/cm
X = 2 ms/cm

Use of the CRO

The previous examples show how the CRO can be used for the specific purpose of measuring amplitude and frequency. There are many other uses to which the CRO can be put. Just consider some of the following:

- Comparing the input and output of an amplifier
- Determining the "ripple" voltage on a rectifier output.
- Inspecting a supply voltage for signs of distortion.

Comparing the input and output of an amplifier

This is best done on a dual-trace CRO.

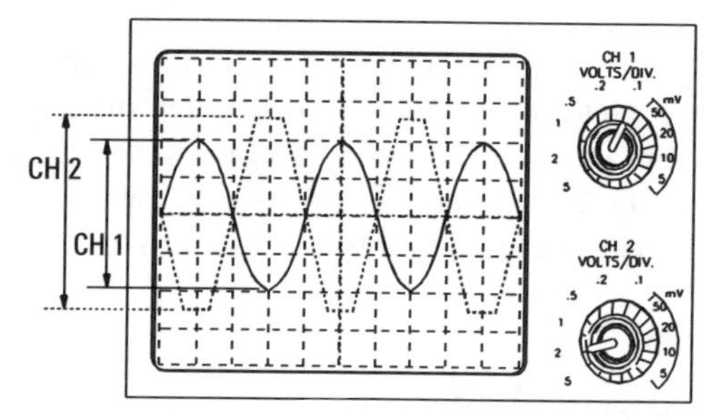

Figure 7.6

Figure 7.6 shows the voltage gain of the amplifier stage. It shows that the output is inverted with respect to the input and that the output is "chopped" and is not a true sine wave. This shows much more information than could have been obtained by meter readings.

Determining the "ripple voltage" on a rectifier output

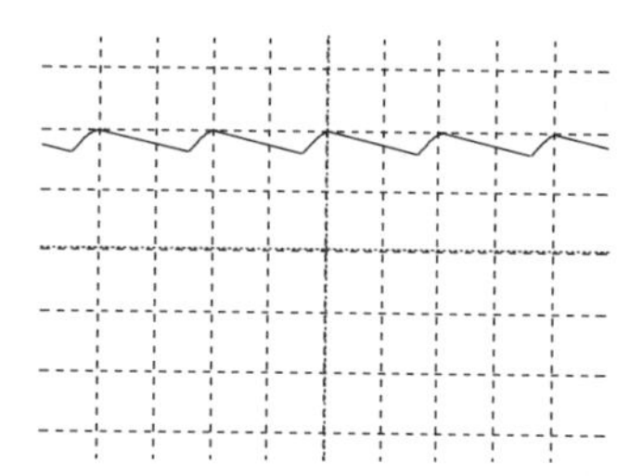

Figure 7.7

It is only necessary to look at the ripple, so this can be amplified by using the Y controls and placed in the centre of the screen by means of the Y shift.

Inspecting a supply voltage for signs of distortion

Figure 7.8 shows how any peculiarities on a sine wave can be seen from the display on the screen.

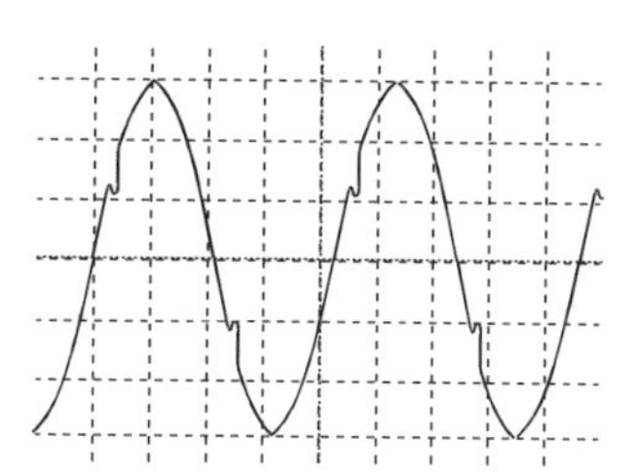

Figure 7.8

Those are just three examples of the use of a CRO. There are many more.

Operation and application of transducers

Any device which changes an input signal in one form into an output signal in another is a TRANSDUCER.

A microphone is an example of a transducer which changes sound waves into an electrical signal. The loudspeaker which changes the electrical signal back into sound is also a transducer. In this section we will look at several types of transducer and consider possible applications for each.

Light activated

Light dependent resistors (LDRs)

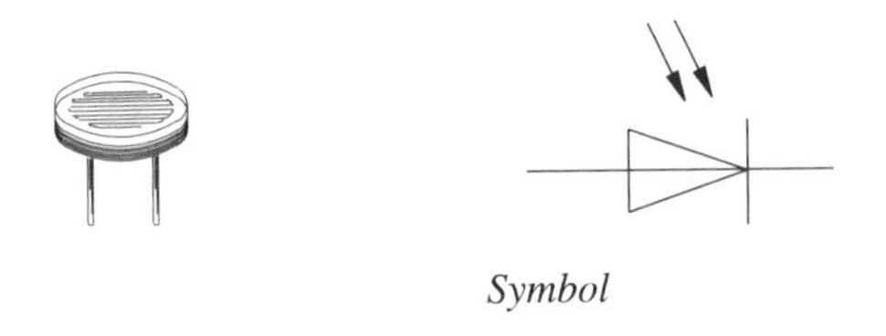

Figure 7.9

LDR's are resistive cells which are sensitive to light. The resistance of the cell decreases when light falls on it. They do not respond very quickly to changes in light level when compared with some other devices but they do handle a reasonable amount of signal current.

This device could be used in a circuit which switches on the lighting automatically if it becomes dark.

Photodiodes

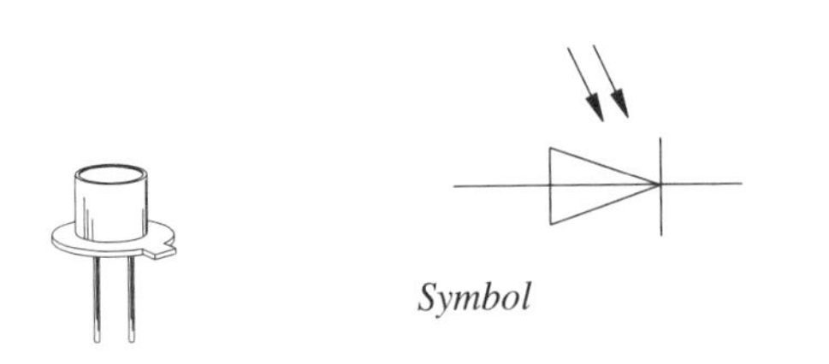

Figure 7.10

These are very fast acting and permit a current to flow which is proportional to the light falling on them but they are only capable of carrying a very small current (5–6 μA)

Photo diodes can be used in communication circuits and in light measuring equipment but the very small signal levels require considerable amplification.

Photo transistors

Phototransistors have clear cases which transmit light. They are able to operate at higher current levels than the photodiode (typically 1 mA). Light falling on the device causes it to become more conductive. It is not as fast as the photodiode but is nevertheless capable of 100 000 operations per second.

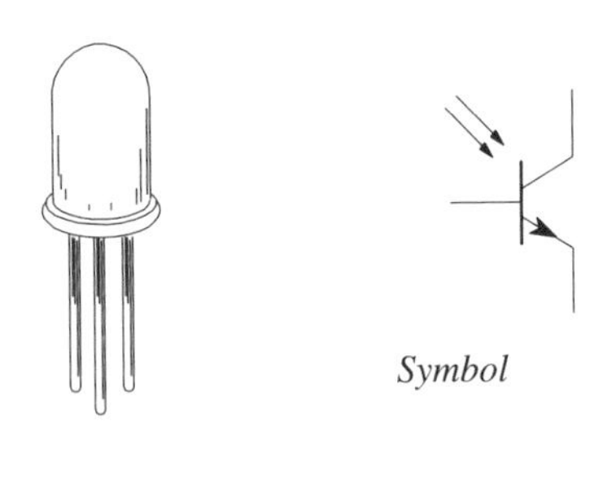

Figure 7.11

Phototransistors are to be found in high-speed counting equipment in the packaging and materials handling industry.

Pressure transducers

Strain gauge

The electrical resistance strain gauge consists of a very thin (less than 1 micron) wire or foil conductor bonded to a flexible insulating backing.

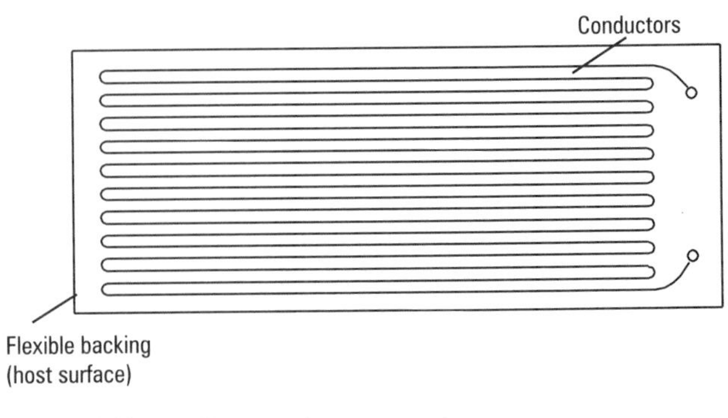

Figure 7.12 *Pressure/stress transducer*

The strain gauge is attached to any surface which is expected to distort under pressure or other mechanical stress.

Any distortion of the host surface which causes an elongation of the conductor will produce a change in the resistance of the device . This is because the stretching effect will increase the length and reduce the cross sectional area of the conductor.

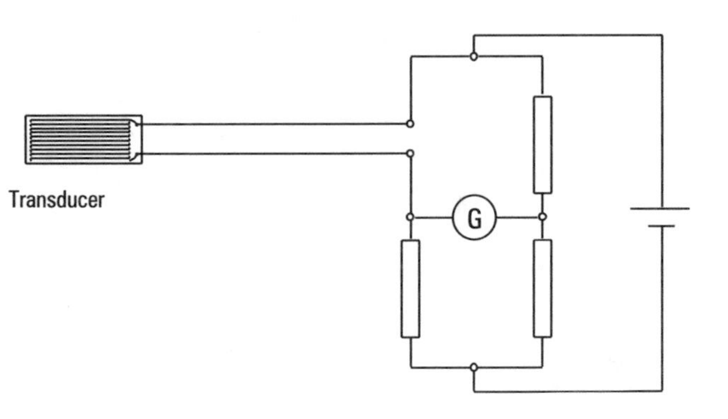

Figure 7.13

The device forms part of a balanced Wheatstone Bridge circuit and any change in resistance will unbalance the bridge thus having a noticeable effect on the circuit.

Piezo-electric pressure transducers

A piezo-electric transducer consists of a thin slice of crystalline material, such as quartz, which is mounted on a suitable backing and cemented to a surface which is to be put under mechanical stress. When the crystal becomes distorted by pressure or tension, it gives off a small voltage which can be detected and amplified into a measurable signal.

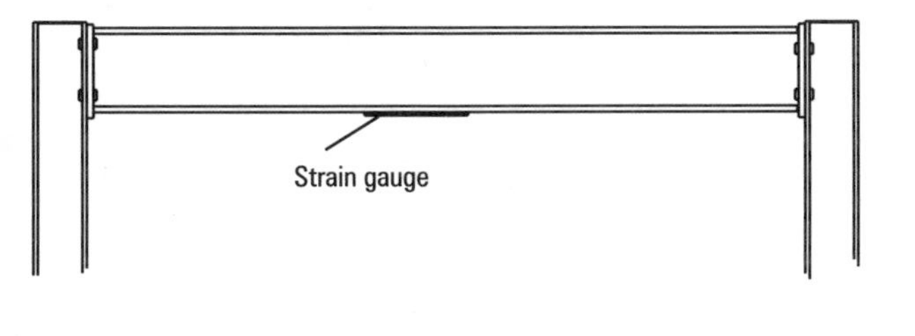

Figure 7.14

Applications of pressure transducers

Pressure transducers are installed in bridges and high-rise buildings in order to measure stresses due to movement in the structure. They are used for scientific research, the nuclear power industry and also in shipbuilding and marine engineering.

The strain gauge can detect very small movements and distortions in metal structures and this is where they are most frequently found but any situation which relies on the measurement of very small deflections could be an application for the strain gauge.

The piezo-electric pressure transducer is also used in the crystal microphone. The pressure of the sound waves on a pressure-sensitive surface is transmitted to the crystal which responds immediately by emitting an electrical signal. The crystal microphone is very small and because it has no electro-magnetic components is also very light which makes it comfortable to use.

Temperature activated

Thermocouples

The thermocouple is a device which, as seen in previous sections, consists of two metal wires welded together to form a junction. If the temperature of this junction is raised an e.m.f. will be produced between this and another junction, the temperature of which has been kept constant. This is known as the "Seebeck Effect" and is widely used in temperature measurement.

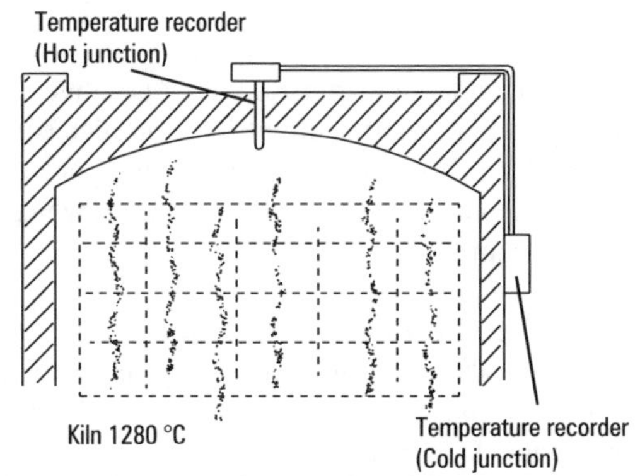

Figure 7.15 *Thermocouple controlling kiln temperature*

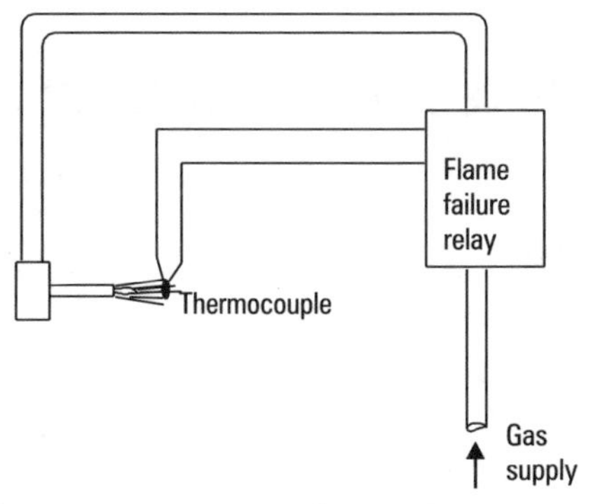

Figure 7.16 *Thermocouple as flame failure detector*

As a transducer, the thermocouple is found to be very reliable provided that the temperature difference is reasonably high. It is widely used as a flame indication device in boilers and other combustion equipment and can be used to provide a signal which will cut off the fuel supply if the flame should go out or fail to ignite.

As a measurement transducer, the thermocouple can be used at temperatures which could not be tolerated by other devices.

Thermistors

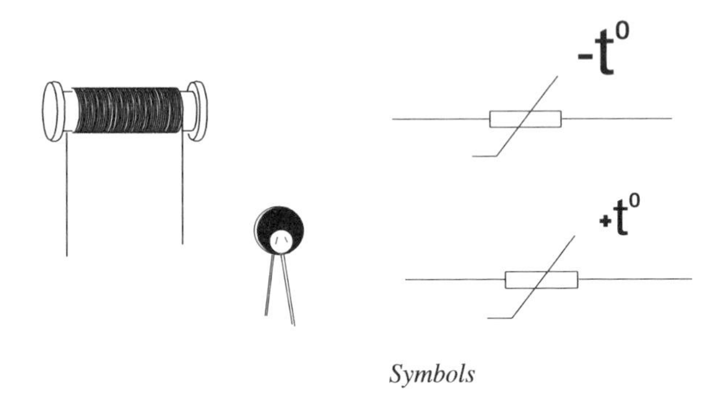

Symbols

Figure 7.17

Thermistors are widely used as heat or temperature transducers. This is due to their special characteristic where the resistance of the device changes quite considerably with the temperature of their environment. They can be selected for positive or negative characteristics according to the type required.

For example, a large thermistor with a negative temperature coefficient will prevent a current surge through equipment on starting. This is because when cold its resistance is high, and this will fall as the device heats up thus allowing a gradual increase in current.

Very small glass-encapsulated bead thermistors may be found embedded in the windings of motors where they can give a very accurate means of determining the actual temperature of the windings. They can form a very important feature of motor protection equipment.

Variable resistors and capacitors

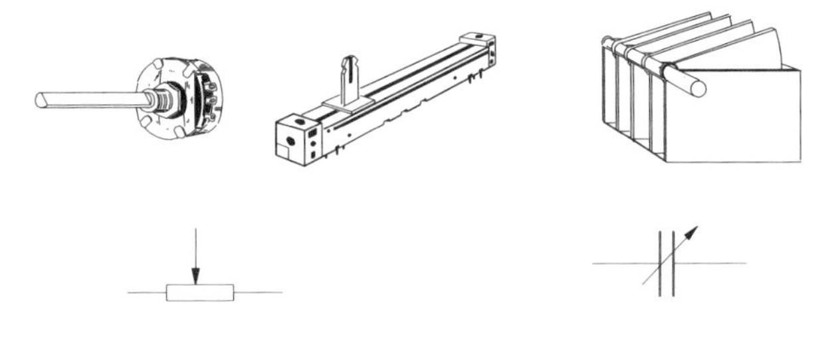

Figure 7.18

These can be used as transducers because the action of turning the spindle, or moving the slide, changes the electrical properties of the device and thus its effect in a circuit.

For example, a change in the height of platform A alters the setting on the variable resistor R. If R forms one limb of a bridge circuit then the balance of the bridge is affected.

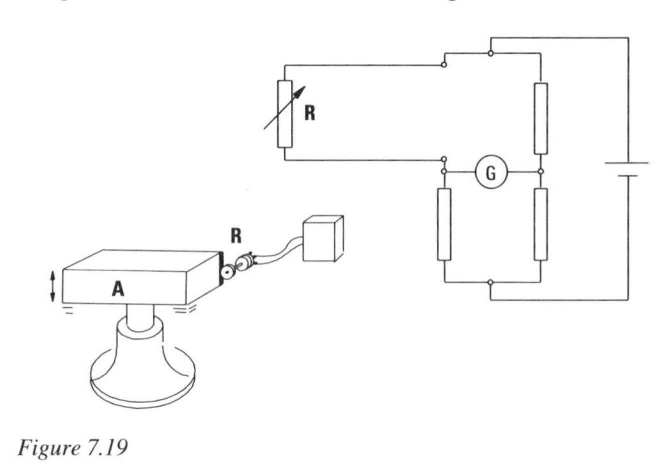

Figure 7.19

Similar applications can be found for the variable capacitor where any rotation can produce a rotation of the spindle. Movement of the spindle controls the capacitance of the device and this can affect the impedance of an a.c. circuit such as the bridge circuit shown in Figure 7.20.

The capacitance of the device can be altered by changing other features of the device such as varying the distance between the plates so that by attaching one plate to a fixed or reference point the relative movement of a moving plate will produce a corresponding variation in the reactance or frequency response.

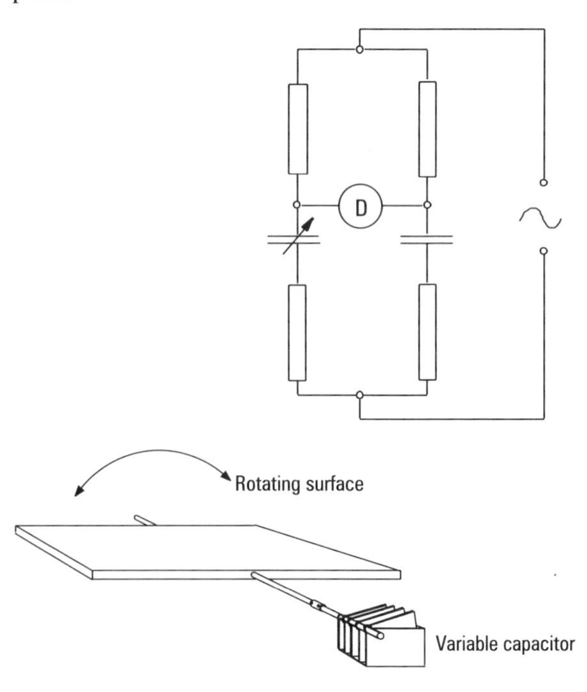

Figure 7.20

Active components

The diode

The principal operational function of the diode is to act as a device which will permit current flow in one direction only.

Of the types of diode in common use there are

- signal diodes
- power diodes
- zener diodes

Signal diodes

Figure 7.21

These are diodes which are small in size and have limited power-handling capability. They are used in high speed switching circuits and high frequency communications circuits.

Power diodes

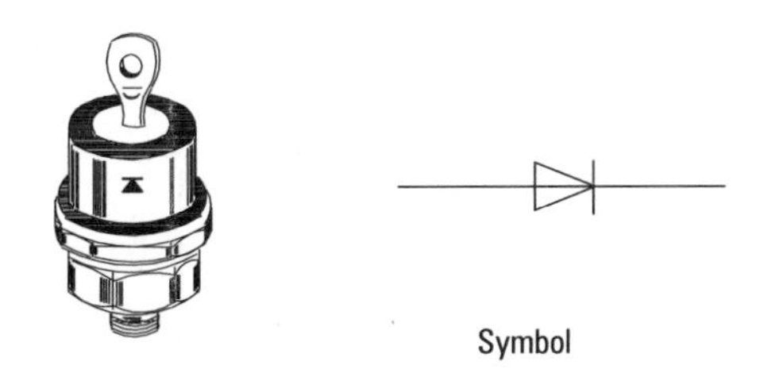

Figure 7.22

Generally these are much larger than signal diodes, and are used for power applications such as mains frequency rectification. They are also used in high speed switching circuits as are found in switched mode power supplies. The stud-mounted diode can be mounted on a heat sink (Figure 7.24) which will help to dissipate the heat generated within the device when it is handling high currents.

Zener diodes

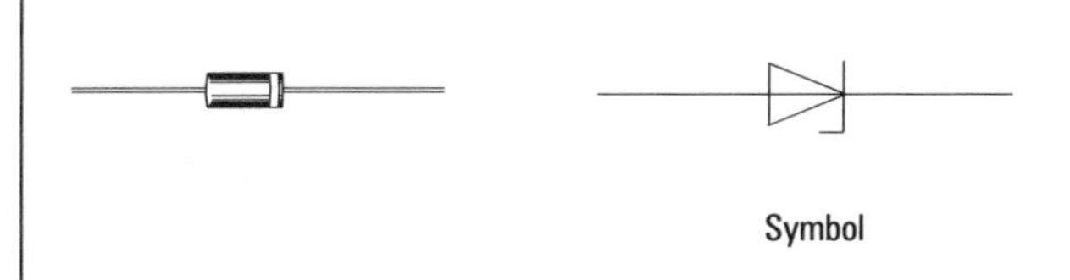

Figure 7.23

Any diode which is reverse biased to a sufficient level will break down and conduct current in the "wrong" direction. When that point is reached, the diode begins to conduct very suddenly and with very little resistance.

A zener diode is a device which makes use of this characteristic by conducting in a reverse biased mode at precisely the same voltage every time that voltage is reached. This makes the zener diode an extremely useful voltage reference device.

An application for the zener diode will be given later on in this section.

Effects of heat on diodes

Reverse leakage current is seldom a problem if the device can be operated within the temperature limits specified by the manufacturer of the device. If a heat sink is used, it must dissipate the unwanted heat effectively otherwise the temperature of the diode will rise and under these conditions the reverse current will increase. An increase in leakage current will cause a further increase in temperature and a situation of thermal runaway will occur which will probably result in the eventual destruction of the component and the malfunction of the equipment.

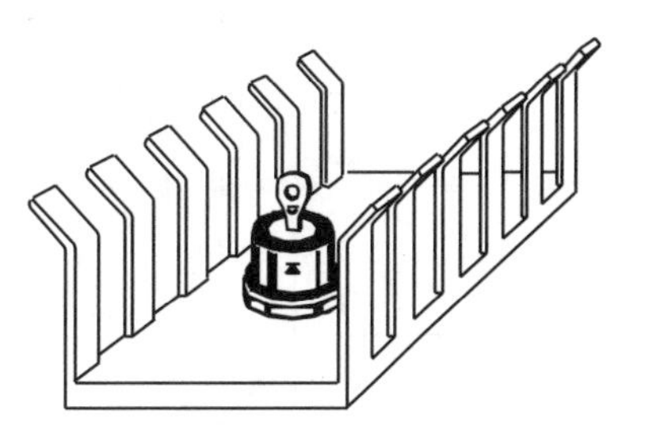

Figure 7.24 Heat sink

Transistors

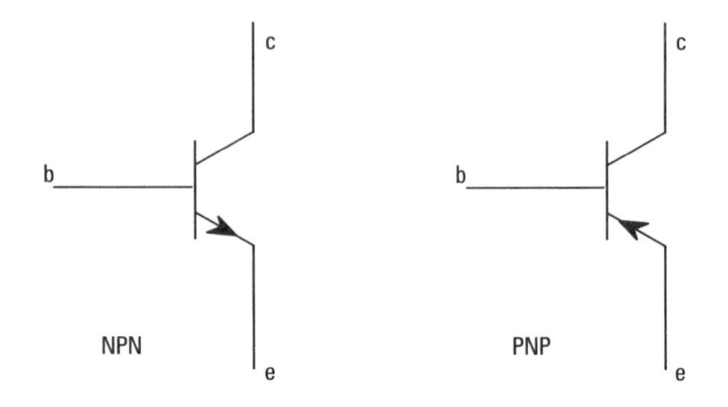

Figure 7.25 *b = base*
c = collector
e = emitter

The transistor is a three terminal device which is capable of amplifying current. The base of the transistor is connected to a potential dividing network which puts a bias voltage in excess of 0.8 V between the base and the emitter. At this stage, the transistor is in a steady state with constant base and collector currents. If the base current is varied by the input of some small signal current then the collector current will also vary. The amount of change in collector current will be greater than the change in base current by a ratio which is known as the forward current gain.

For example, if the base current is increased by 0.1 mA and this produces a corresponding change in collector current of 10 mA then the current gain is

$$\frac{10}{0.1} = 100$$

This can be sued as a switching function, where a signal current of relatively small proportions can be fed into the base and this will produce a larger current in the collector circuit which can be used for a higher power application. The collector current will only flow as long as the base current is present.

Figure 7.26 shows a circuit where a small input signal controls the operating coil of a relay.

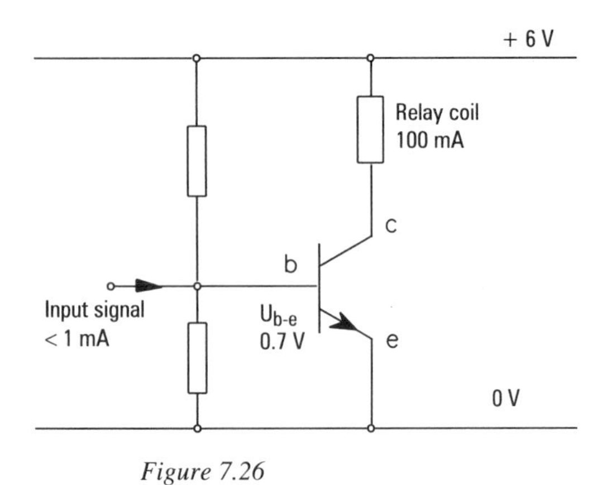

Figure 7.26

Thyristors

The thyristor, or silicon controlled rectifier (SCR), is a device which is used in switching circuits.

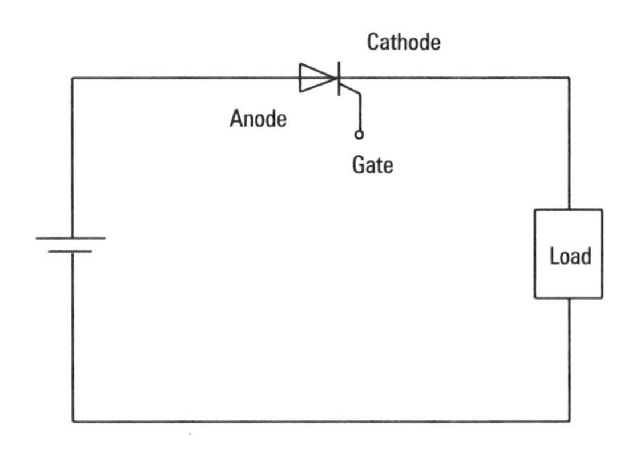

Figure 7.27

This is a four-layer device which does not conduct in either direction when first connected to the circuit.

If, however, a short pulse is applied to the gate it will begin to conduct and will continue to conduct after the gate pulse has been removed.

The thyristor is a d.c. device and will only conduct in the direction indicated. If connected to an a.c. supply it will cease conducting after the supply current has reached zero and will only switch on again if the gate pulse is re-introduced in the next or some subsequent positive half cycle.

The circuit shown in Figure 7.28 includes two thyristors, SCR1 and SCR2, connected "back-to-back". This then allows for connection to a.c. circuits so that one of the thyristors will operate on each half cycle.

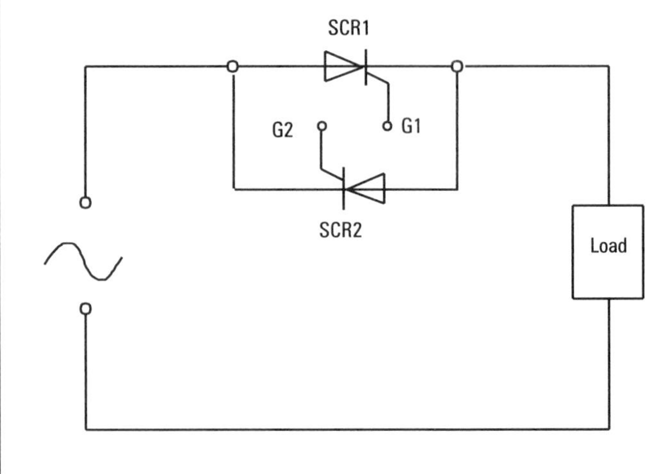

Figure 7.28

Diacs

These are two terminal devices which conduct in either direction but only after the voltage across them has reached the level required for conduction to take place. They can be regarded as "back to back" zener diodes.

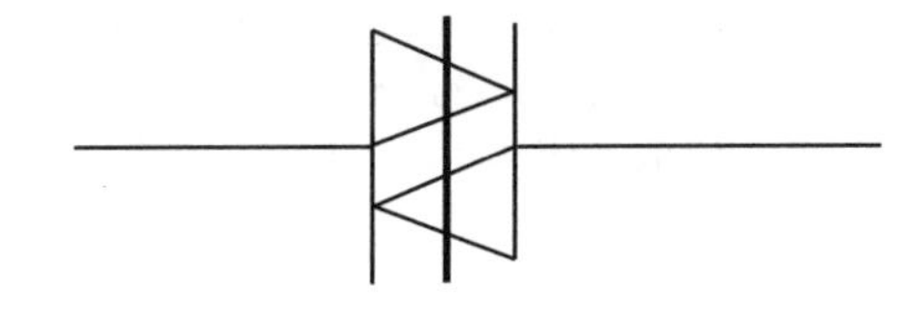

Figure 7.29 *Diac symbol*

The normal function of the diac is as a triggering device for a thyristor or triac so that the device switches on when the signal level has reached a predetermined value.

Triacs

The triac is a bi-directional switching device which can be used in a.c. circuits.

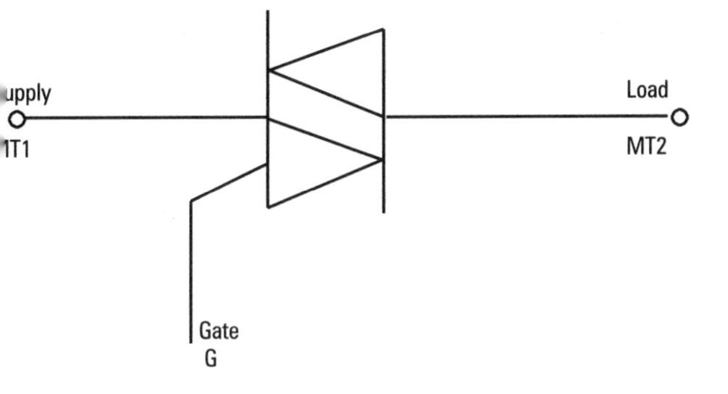

Figure 7.30

The main terminals are identified as MT1 and MT2 and are connected as shown in Figure 7.30. The triggering impulse is received at the gate (or G terminal). The triac is only used for comparatively low power applications such as lamp dimmers. For heavy power applications, up to several hundred amps, "back to back" pairs of SCRs will perform the same function.

Component recognition

Discrete active electronic components can, in the majority of cases, be recognised by certain features of their construction.

Small power and signal diodes will be recognisable by the presence of a colour band or indentation at the cathode, or receiving end, of the component.

Larger power diodes with radial leads may be bullet shaped with the rounded end pointing towards the direction of current flow.

Stud-mounting diodes must be treated with some care as they can be packaged either way round. Check with the manufacturer's data sheet, or test with an ohmmeter, before putting one into a circuit, otherwise you may cause damage to other components. It is often the practice of manufacturers to print the diode symbol on the side of the device and in this case current flow will be as indicated by the point of the symbol.

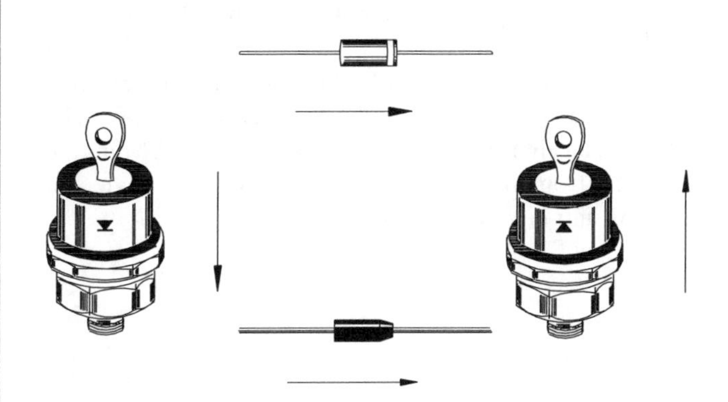

Figure 7.31 *Arrow shows direction of current flow*

Transistors come in a variety of packages and there are many variations such as small can, large can, plastic encapsulated, flat pack and high power metallic enclosures. There can be a variety of lead arrangements within each of the types, and reference to the manufacturer's data sheet for each component is the only reliable way of identifying the terminals.

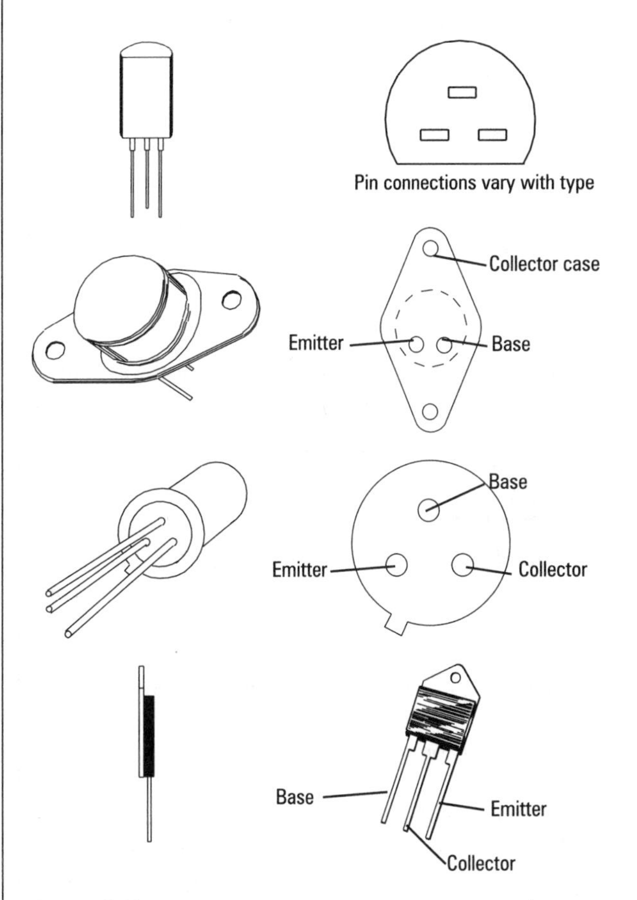

Figure 7.32

Integrated circuits (ICs)

The integrated circuit is a complete electronic circuit formed from one piece of semiconductor material. The active component may consist of several transistors along with their associated resistors, diodes and capacitors all incorporated in a tiny chip of silicon and encased in a plastic package.

For example, an operational amplifier of the 741 series is a device which has an open loop gain of many hundreds of thousands which is modified by the connection of external resistors to any value required by the user, (say 1000). The package is an eight-pin dual in-line assembly and the pins are numbered from the left of the notch looking down and following anti-clockwise round the component.

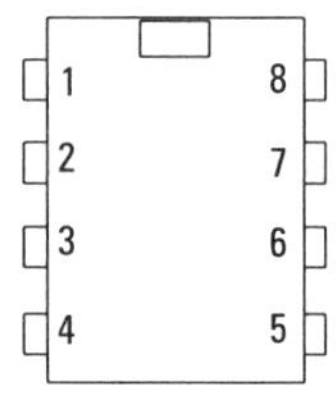

Figure 7.33

A 14-pin dual in-line package may have a dot next to pin No 1 and the following pins numbered as before.

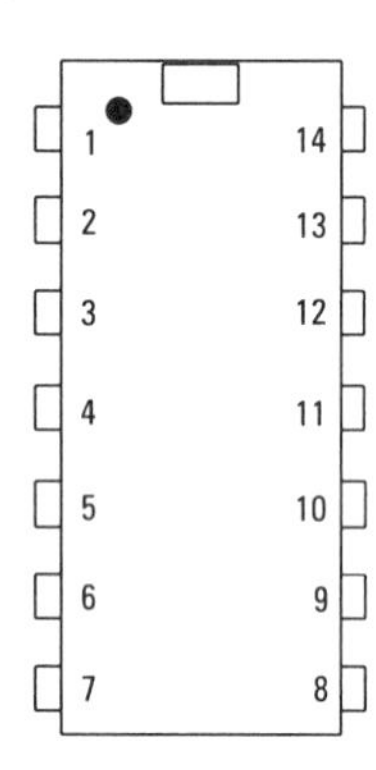

Figure 7.34

No matter how many pins a package has, and it may be as many as forty, the numbering system is the same.

Printed circuit boards

A printed circuit board is an insulated base with tracks of conducting material running in an intricate pattern over one or both sides of the board.

The board may start life as a sheet of copper-faced laminate from which the unwanted parts are removed by chemical etching or by a precision milling process.

The main advantage of the printed circuit board (PCB) is that once the design is complete and a master model has been made, an infinite number of copies can be produced from the same pattern.

Handling printed circuit boards should always be carried out with care. The manufacturing process of the board means that they may be very thin in places and can crack easily. Consideration must also be given to the fact that many of the components on the boards may be damaged by the voltages in the human body. So when removing and replacing printed circuit boards they must be disconnected by unplugging leads or the boards removed from their edge connectors. The boards must be handled by their edges without touching the conductor tracks. It may be necessary to wear an earthed strap to stop the effects of static electricity from damaging the sensitive components.

Circuit components or their mounting bases are connected directly onto the board by soldering.

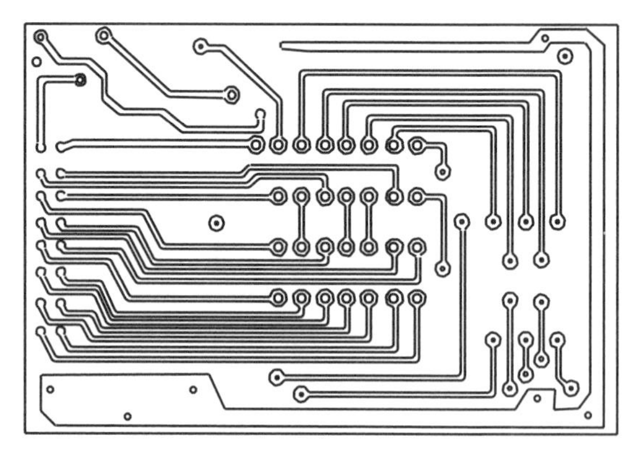

Figure 7.35 *Printed circuit board*

Great care must also be exercised when replacing the components which have been mounted on a PCB. The copper tracks are very fine and can be fractured if the board is bent or flexed during the process.

If excessive heat is applied when de-soldering, the tracks can become detached from the board and broken. In addition to this, the components themselves can be damaged by heat.

In order to avoid damage try to follow some basic rules.

1. Cut away the component to be discarded before de-soldering.
2. Use a small soldering iron and "solder sucker" or de-soldering wick to clear the cut ends away.
3. Clear the holes in the board and mount the new component making sure that everything is perfectly clean.
4. Attach a heat shunt to the component leads to prevent damage during the soldering process.
5. Use a fine tipped soldering iron suitable for PCB work. Solder as quickly as possible using 60/40 solder with a resin core. Do not use any more solder than is necessary to make a joint which is electrically sound. Do not move the component until the solder has solidified.

Remember that these are basic rules only. Your lecturer or instructor may have others which you should observe - follow the advice that you are given.

Remember

The fumes given off when soldering contain lead!

D.C. power supplies

A circuit which can convert a conventional a.c. supply into a usable d.c. output is what is commonly described as a "power supply". It is a collection of components which will consist of

- a transformer to set the required level of output voltage and, if necessary, provide isolation from the mains supply
- a rectifier to convert the a.c. output of the transformer into a uni-directional current.

If further sophistication is required it may also have

- a filter or smoothing stage which will take the" humps" out of the rectified output and provide a smooth d.c. and
- a stabilising circuit which will give the power supply a constant voltage output under varying load conditions.

The simplest form of power supply is a half-wave rectifier which uses only one diode.

Half wave rectifier

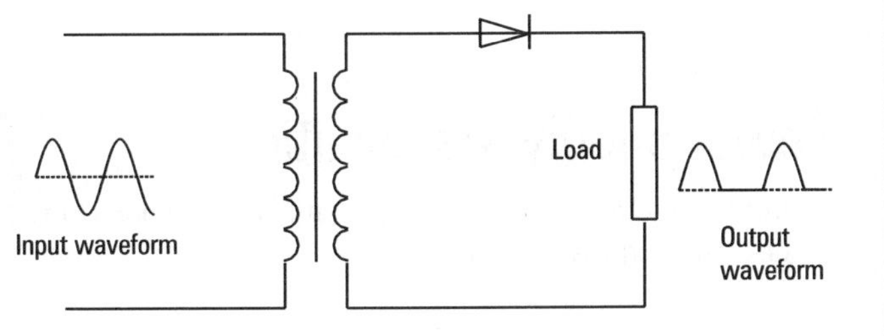

Figure 7.36

As can be seen from the output waveform, this is only capable of delivering an output from alternate half cycles of the supply. This is a very basic circuit and can only be used where the quality of the output is not an important factor. It could be used for battery charging, but would not be considered suitable for smoothing to give an eventual pure d.c. output.

Full wave bi-phase rectifier

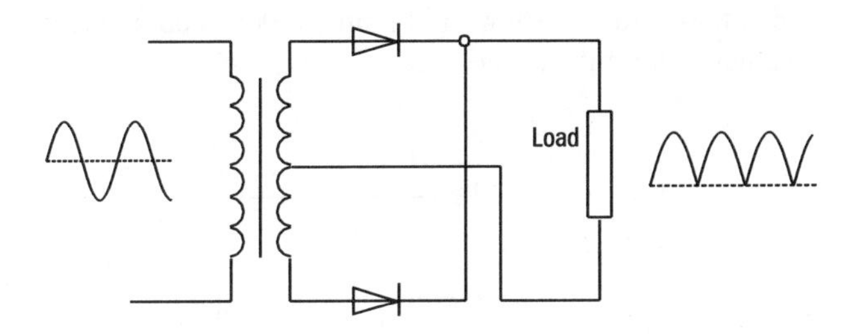

Figure 7.37

This circuit uses a centre-tapped transformer and two diodes to give a rectified output on both half cycles of the a.c. supply.

This type of power supply can be found in electronic equipment and it is a good source of d.c. The problem is that only half of the transformer secondary is in use at any one time and so it has to be bigger than is strictly necessary for the power output required.

Bridge rectifier

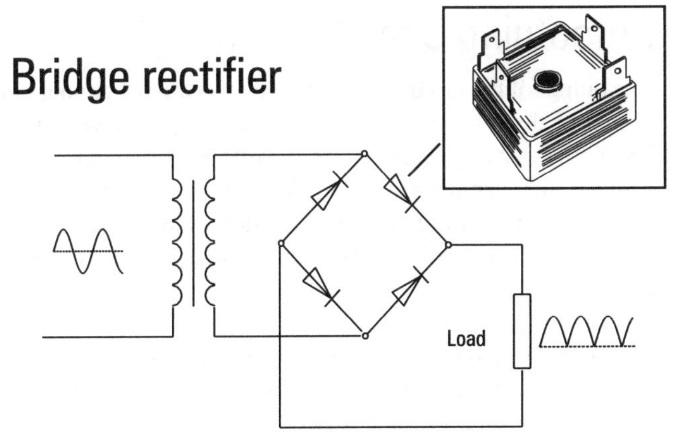

Figure 7.38

This is by far the most popular type of rectifier. It has a full wave output which is delivered from four diodes. The diodes can be discrete components or sealed in a purpose-made package. The transformer secondary is fully utilised and the output can be smoothed and stabilised if a pure d.c. is required.

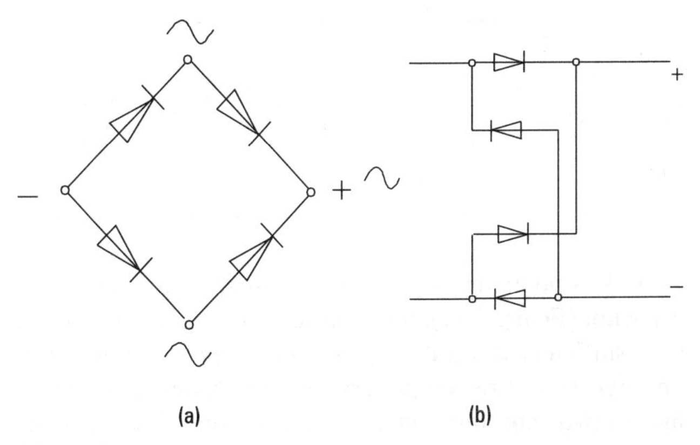

Figure 7.39

The arrangement of diodes shown in Figure 7.39(a) is generally used for the circuit diagram but if laid out on a circuit board, the lozenge shape is a waste of space so the diodes are laid side-by-side, as shown in Figure 7.39(b), and connected together to form the bridge.

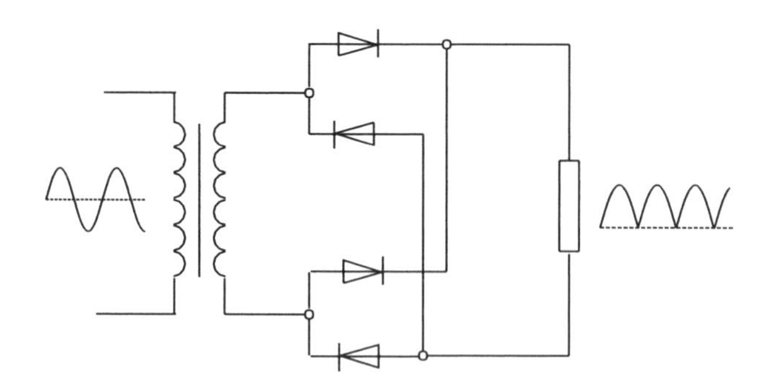

Figure 7.40 *Full bridge circuit using the "side by side" method of connection*

Smoothing circuits

The output of the rectifier circuits shown is unidirectional but cannot be described as pure d.c. This is because the voltage is constantly varying between zero volts and the maximum value. If, for example, any of these circuits were used to supply the amplifier of a public address system, the resulting hum created by the unsmoothed d.c. would drown out any attempts at producing intelligible speech.

A smoothing circuit acts as a filter which has the effect of filtering out the unwanted rise and fall in output voltage leaving a d.c. output which is free from ripple.

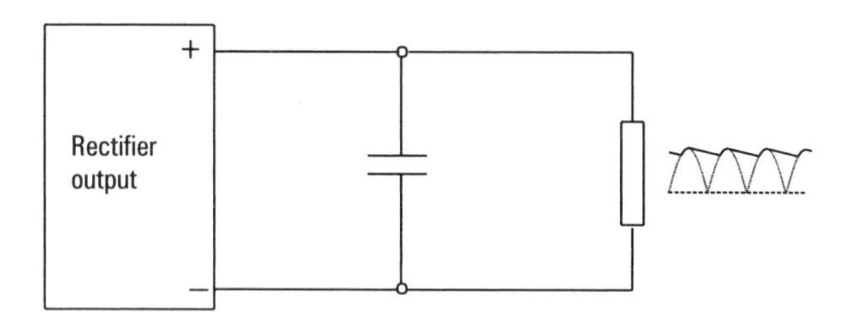

Figure 7.41

A basic smoothing circuit could consist of a single large capacitor (Figure 7.41). If the capacitor is large enough it will store sufficient charge to keep the power supply output voltage up when the rectifier output drops to zero. A smoothing circuit having two capacitors and a resistor is more effective than a single capacitor (Figure 7.42).

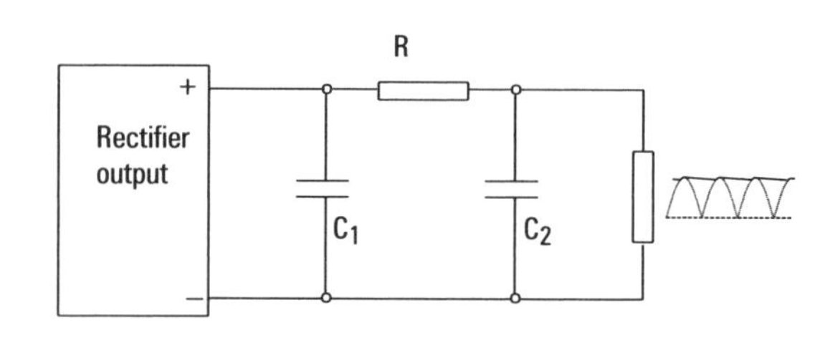

Figure 7.42 *R - C filter*

This is known as an R–C filter. It is excellent for light loads. Most of the ripple, which remains after C1, is attenuated by the resistor. For higher load currents the resistor would create problems with voltage drop.

For heavier loads an L–C filter would be more suitable as this relies on the reactance of the inductor. The inductor impedes the flow of the ripple current but allows the d.c. to pass unhindered. It would be more expensive than the R–C filter because of the relatively high cost of the inductor but it is effective in reducing ripple with little loss of output.

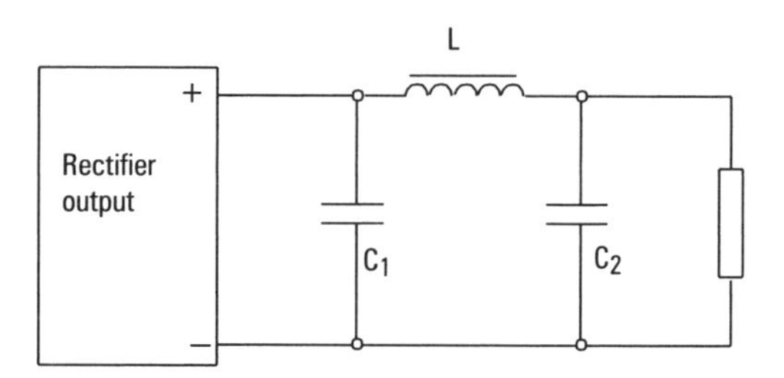

Figure 7.43 *L–C filter*

Power supply stabilisation

A simple but effective stabilising circuit can be made using a zener diode and resistor.

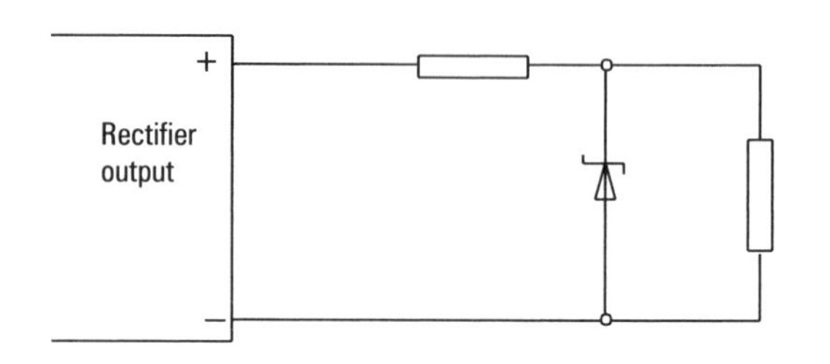

Figure 7.44

The zener diode will conduct when the reverse bias voltage reaches a pre-determined value. The resistance of the zener in this condition is very low and any increase in voltage will result in an increase in current through the device.

For example:

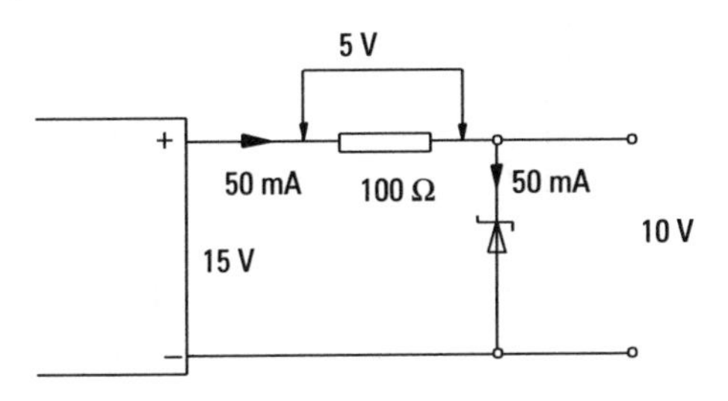

Figure 7.45

The output of the power supply is 15 V and the zener diode conducts at a voltage of 10 V. The voltage across the 100 W resistor is therefore 5 V which means that the current through the resistor is 50 mA. As there is no load, the current through the zener is 50 mA.

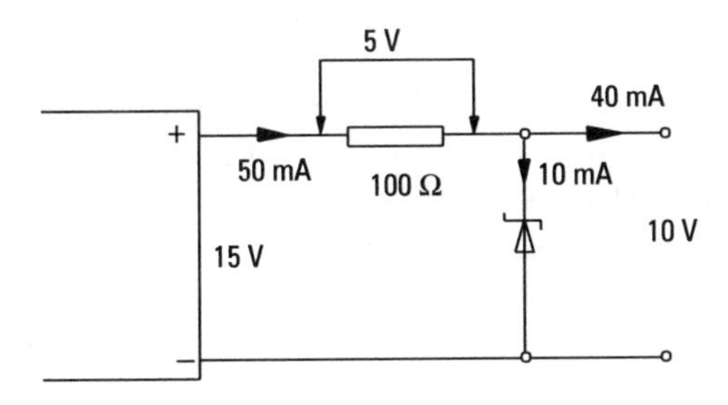

Figure 7.46

At a load current of 40 mA the zener is still conducting. Since the zener voltage is 10 V, the current through the 100 W resistor is still 50 mA but now 40 mA is going into the load and only 10mA through the zener. As long as the load current of this circuit does not exceed 50 mA the zener diode will stabilise the output.

If the output current exceeds 50 mA, the voltage drop across the 100 W resistor will be more than 5 V, the zener will switch off because its threshold voltage has not been reached and the circuit will no longer be stabilised.

Try this

Draw the circuit diagram of a simple zener diode stabilising circuit and calculate the value of resistance required to stabilise a 16 V rectifier output to 12 V at load currents up to 500 mA.

Amplifiers

The amplifier is a device which takes an input signal at a low level and produces an output signal identical to the input but at a higher level.

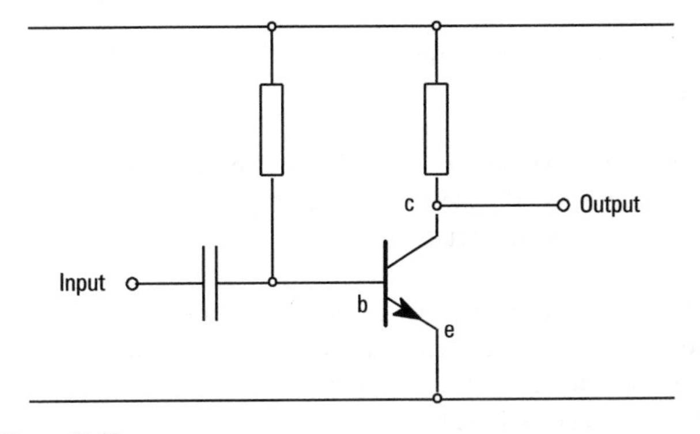

Figure 7.47

A common-emitter transistor amplifier, such as the one shown in Figure 7.47, uses the current amplification effect of the transistor to produce an output voltage by varying the current through the collector load resistor.

The steady state base current is influenced by an oscillating signal voltage.

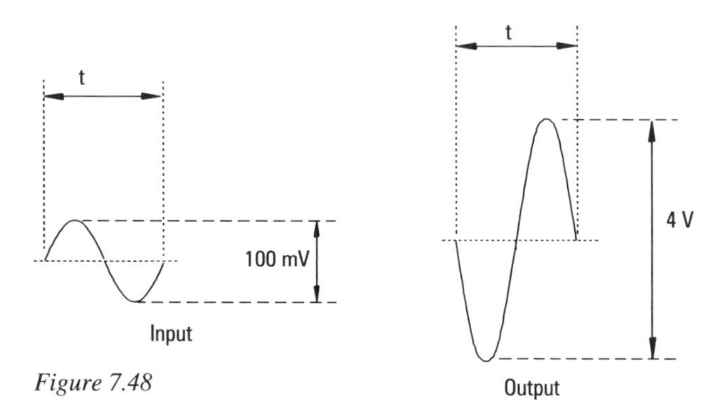

Figure 7.48

Any change in base current will produce an immediate change in collector current, but many times greater. If the base current increases, the collector current increases but more so. The increased collector current causes an increased voltage drop in the collector load resistor. So, as the base goes up, the collector comes down and vice versa. An alternating voltage applied to the base will produce an alternating voltage at the collector, but inverted, as shown in Figure 7.48.

The difference in amplitude between the input voltage and the output is termed the "gain" of the amplifier.

For example, if an input signal voltage of 100 mV peak to peak produces an output of 4 V peak to peak then the gain of the amplifier is 40.

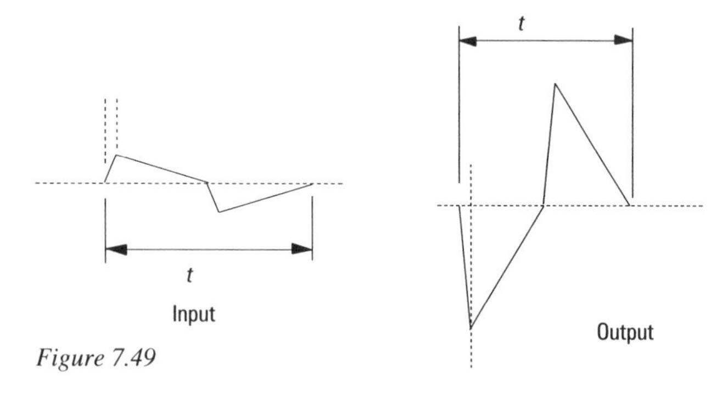

Figure 7.49

This is an inverted output but it is otherwise identical to the input.

Operational Amplifiers such as the 741 have a very large gain of about 100000. They also have the facility for providing inverted or non-inverted signals.

This is very important in the case of operational amplifiers because a very large proportion of the output is fed back into the input as an inverted signal. This is what is known as negative feedback and is used to reduce the overall gain and improve the stability of the circuit.

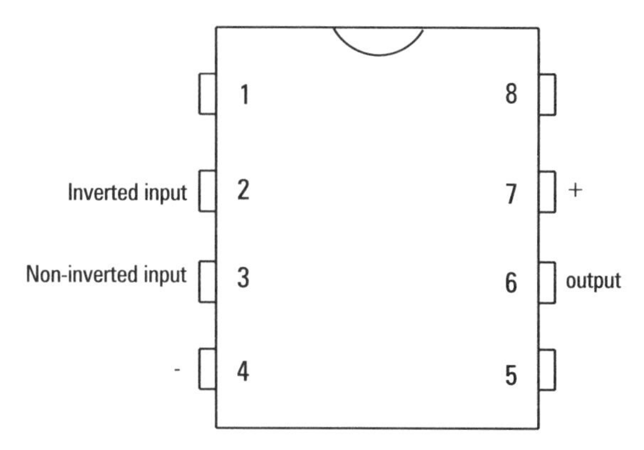

Figure 7.50

Signal sources

It would be virtually impossible to test electronic circuits without some form of signal generator.

The most basic of signal generators will provide a signal as an aid to testing or fault diagnosis varying in frequency from a few tens of Hertz to several megahertz. The same equipment will also provide signal voltages from several volts right down to microvolts. With this and an oscilloscope, the electronic engineer can investigate the workings of most circuits in order to carry out tests and repairs.

Although the sine wave is most frequently used, many instruments will be able to deliver square wave, saw-tooth or modulated output.

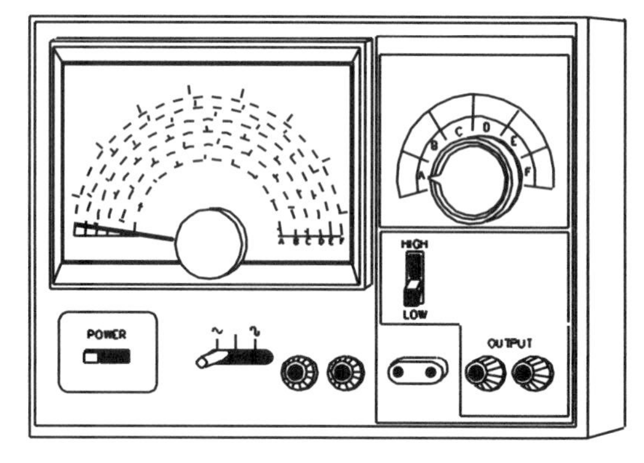

Figure 7.51 *Signal generator*

Digital signals

Digital signals are essentially a series of pulses which indicate either one of two states; ON or OFF. The ON state signifies a binary 1 and the OFF state a binary 0. The sequence and grouping of the pulses can be encoded and decoded, thus converted from analogue to digital and back again as required.

An analogue system has to respond faithfully to an infinite number of signal levels in order to transmit information accurately. This makes it vulnerable to unspecified errors. A binary digital system only has to respond to two states, either 1 or 0 and since these levels can be set within wide tolerance limits the equipment is greatly simplified and much more reliable.

To investigate the behaviour of digital equipment a pulse generator is an essential piece of test equipment. Since the logic state can be either positive or negative the device must be capable of delivering one or the other as required. An internal generator generates square wave pulses of up to 18 V amplitude at a constant frequency of 1 kHz and a selector switch will enable the operator to inject a single pulse, a pre-determined number of pulses or a continuous output. The logic probe is a partner to this device as it is used to detect and give a clear indication of the output state of the component under test.

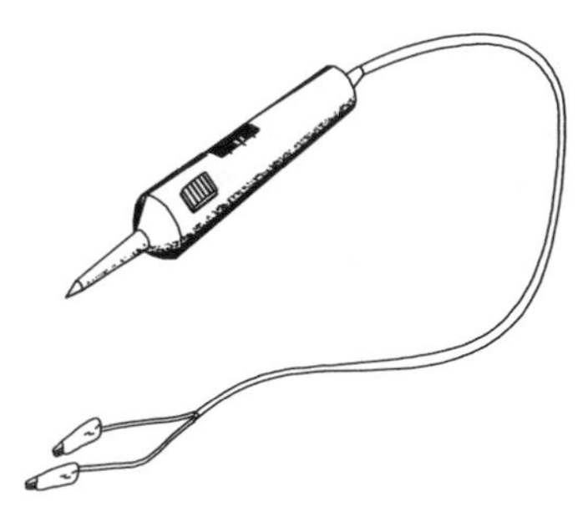

Figure 7.52 *Pulse generator*

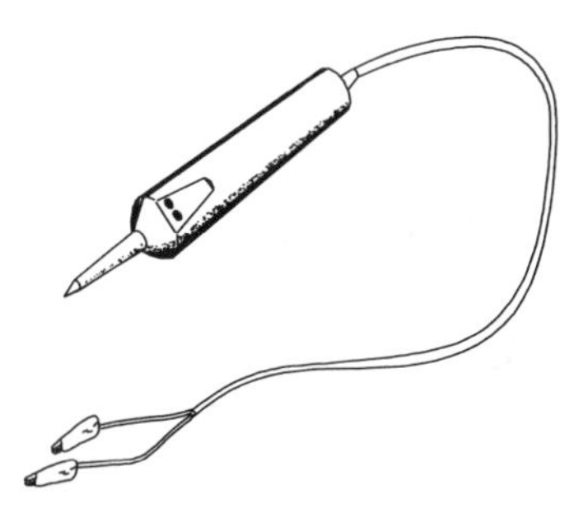

Figure 7.53 *Logic probe*

Logic gates

Although most of us have appreciated the excellence of digital recording and sound reproduction equipment, it may not be generally realised that digital techniques have also revolutionised industrial and environmental control processes.

Logic gates are capable of processing digital signals in order to give an output signal when the appropriate input conditions are present.

In basic form there are three different types of gate:

- AND
- OR
- NOT

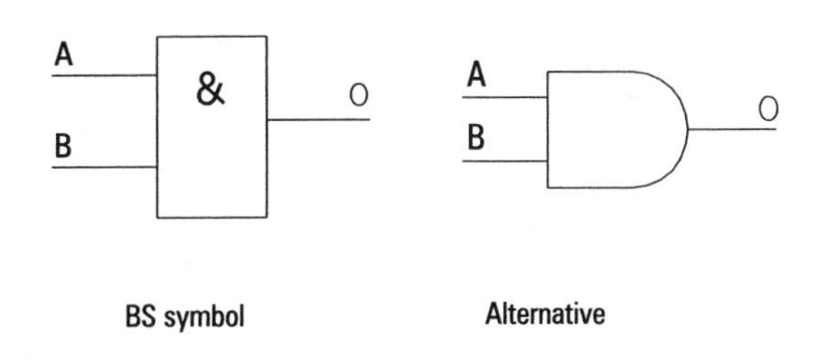

Figure 7.54 *AND gate*

An **AND** gate gives an output when an input is present at all, (in this case both), the inputs. (Both A & B must be ON.)

The truth table shows the state of the output for all input states:

Switch A	Switch B	Output	
A	**B**	**0**	
0	0	0	A OFF & B OFF = No output
0	1	0	A OFF & B ON = No output
1	0	0	A ON & B OFF = No output
1	1	1	A ON & B ON = Output

A & B = 1

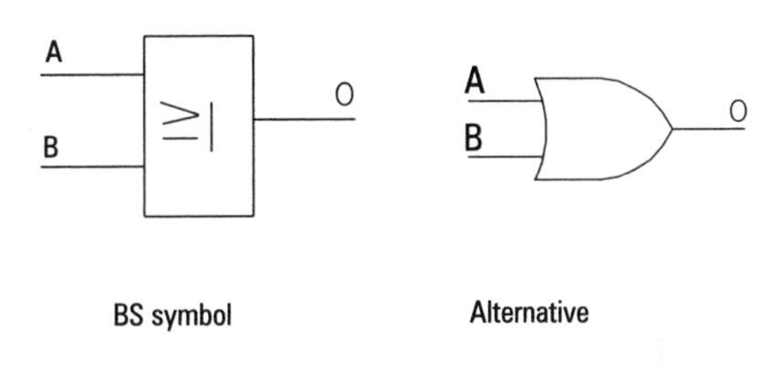

Figure 7.55 *OR gate*

The **OR** gate gives an output when a signal is present at one or the other input

A	B	0
0	0	0
0	1	1
1	0	1
1	1	1

A or B = 1

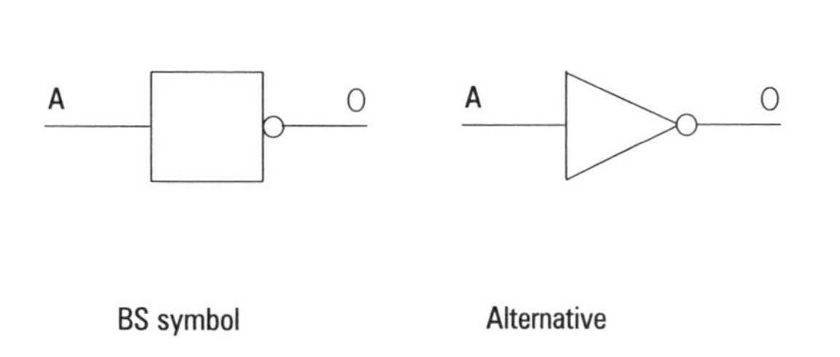

Figure 7.56 NOT gate

The **NOT** gate is an inverter which changes the state of the input.

In other words it gives an output when there is **NOT** an input.

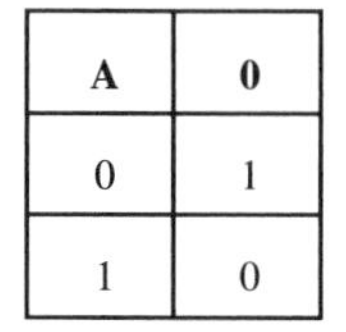

A	0
0	1
1	0

The **NOT** gate can be incorporated with either of the other two to make a **NOR** (not or) or a **NAND** (not and)

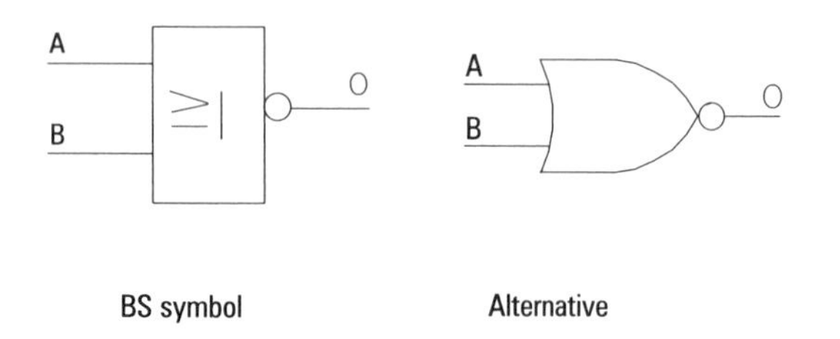

Figure 7.57 NOR gate

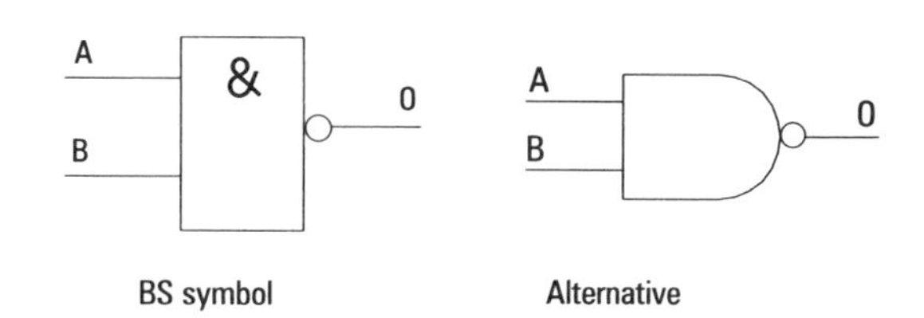

Figure 7.58 NAND gate

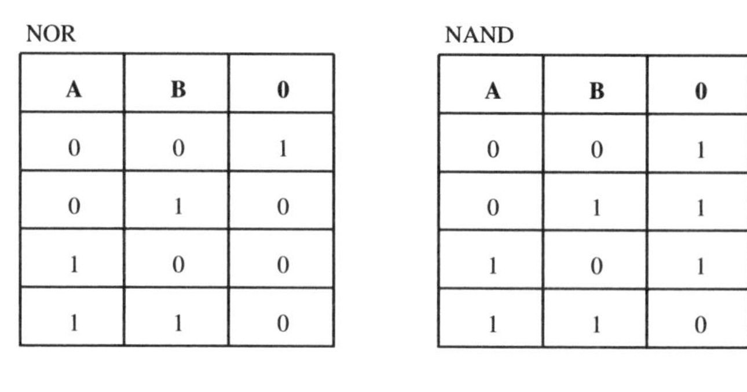

NOR

A	B	0
0	0	1
0	1	0
1	0	0
1	1	0

NAND

A	B	0
0	0	1
0	1	1
1	0	1
1	1	0

NOT (A OR B) **NOT (A & B)**

Logic gates can be combined to provide a control function which will deliver one or more output signals when certain input conditions are present.

Follow the truth table for this four input circuit

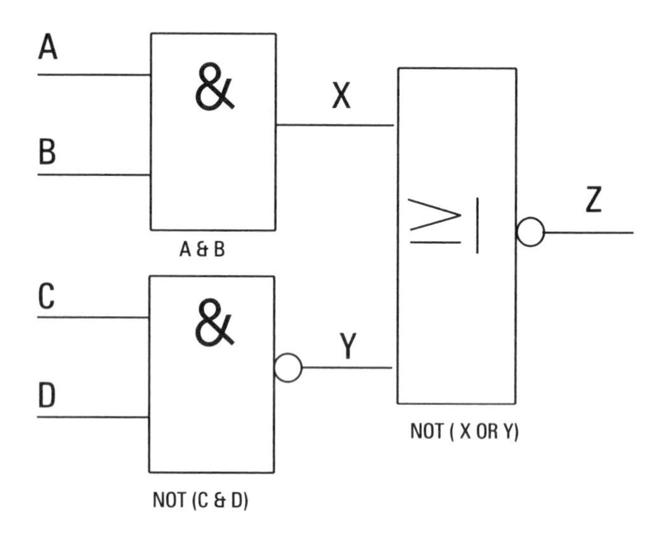

Figure 7.59

For a four-input circuit there are sixteen possible states.

A	B	C	D	X	Y	Z
0	0	0	0	0	1	0
0	0	0	1	0	1	0
0	0	1	0	0	1	0
0	0	1	1	0	0	1*
0	1	0	0	0	1	0
0	1	0	1	0	1	0
0	1	1	0	0	1	0
0	1	1	1	0	0	1*
1	0	0	0	0	1	0
1	0	0	1	0	1	0
1	0	1	0	0	1	0
1	0	1	1	0	0	1*
1	1	0	0	1	1	0
1	1	0	1	1	1	0
1	1	1	0	1	1	0
1	1	1	1	1	0	0

From the truth table you will see that there are only three possible states of the four inputs which will result in an output.

Try this

Draw up a truth table showing the output condition for all inputs.

Thyristor control circuits

You will have learned from the earlier section dealing with thyristors, that the device can be switched on by means of a short pulse of d.c. at the gate. This action is described as triggering or firing the thyristor. Once it has been fired, the thyristor continues to conduct until the cathode voltage is brought to zero. The the precision with which the trigger circuit times the injection of successive pulses is crucial to the correct operation of the circuit.

It will not be possible to investigate the circuitry which lies behind the trigger circuit but the basic principles will be discussed.

Amplitude control

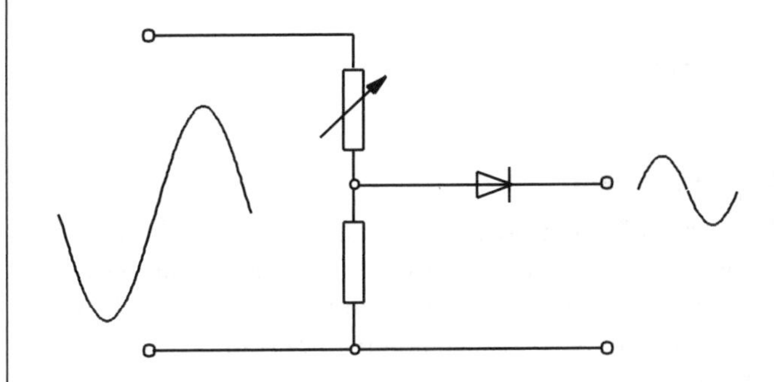

Figure 7.60

The signal voltage being used to trigger the thyristor is a proportion of the mains voltage used to supply the load. The time taken for the trigger voltage to fire the device will depend on the amplitude of the controlling wave.

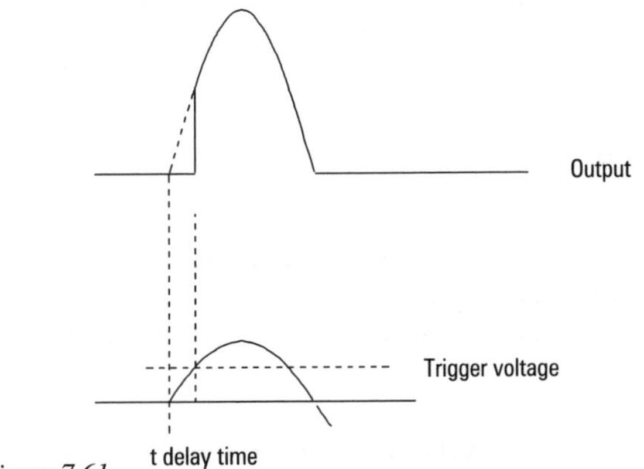

Figure 7.61

Because there is no phase shift in the controlling wave, it can only trigger the thyristor during the first quarter of the cycle.

Phase control

By introducing phase shift into the triggering circuit, the onset of the pulse can be delayed beyond the maximum value of the mains waveform.

Phase angle control could be achieved by means of an R–C circuit which would delay the rise in trigger voltage.

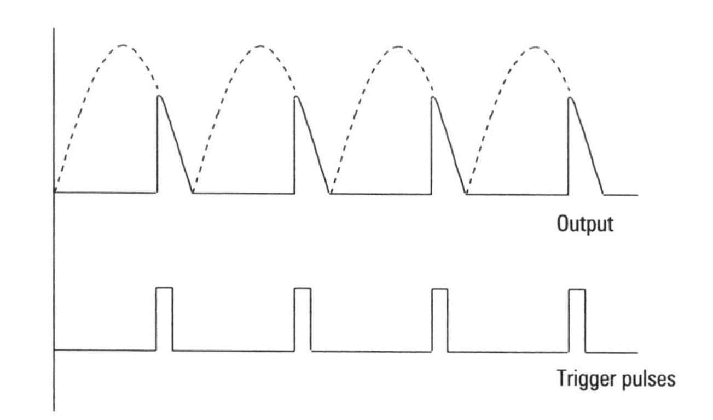

Figure 7.62

Burst control

In the case of heating loads and similar applications it may not be necessary to fire the thyristor at the same instant in every cycle. It may be quite sufficient to turn the supply on and off in bursts of many seconds duration by means of extended gate signals.

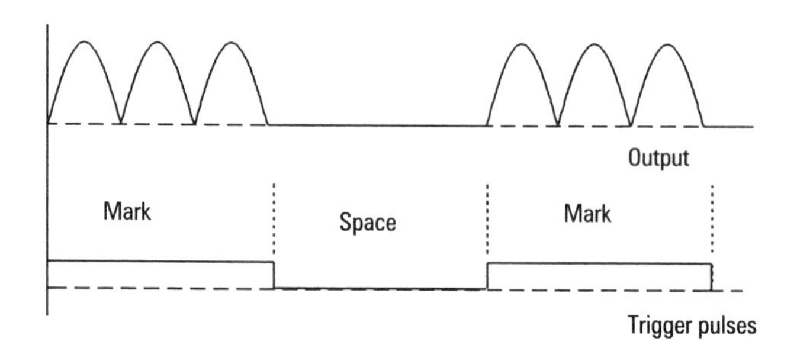

Figure 7.63

The heat output is then controlled by the relative length of the periods when the supply is on compared to that when it is off. The ratio of on to off periods is an example of " mark-space ratio".

When they are of equal length the mark-space ratio is 1:1 and the energy throughput is halved.

For most heavy current a.c. thyristor applications, two thyristors would be operated in conjunction, one for each half cycle of the supply. they would normally operate independently of each other but if controlled from a common source such as a pulse generator then there would need to be some form of isolation such as a pulse transformer.

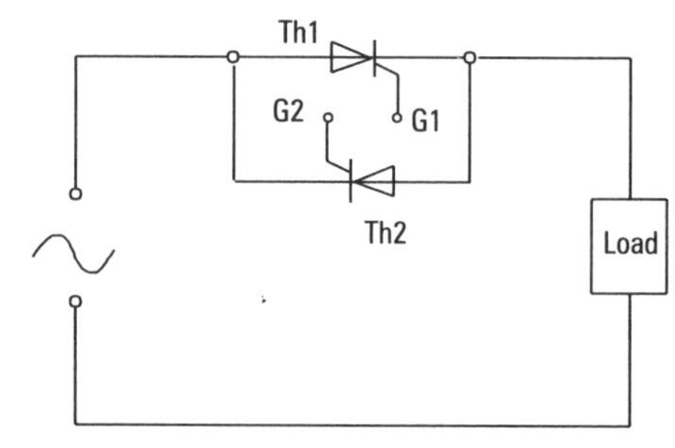

Figure 7.64

Use of the pulse transformer

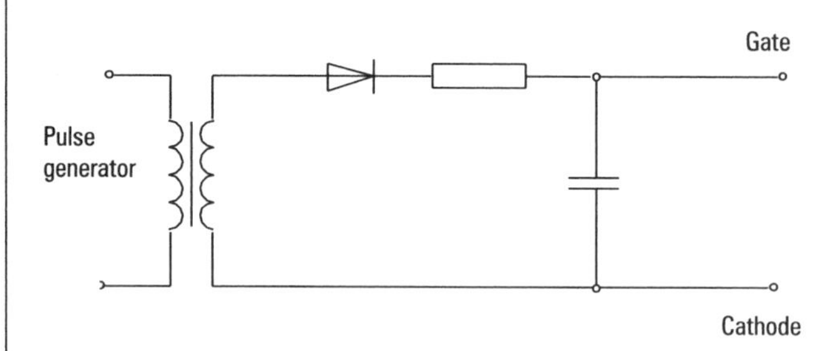

Figure 7.65 *The triac*

Three-phase applications

Thyristors are used in three phase circuits and can be found in power supplies and motor speed control equipment.

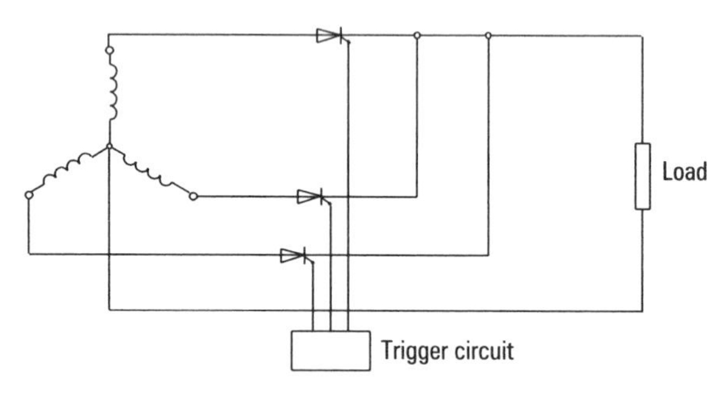

Figure 7.66 *Power supply*

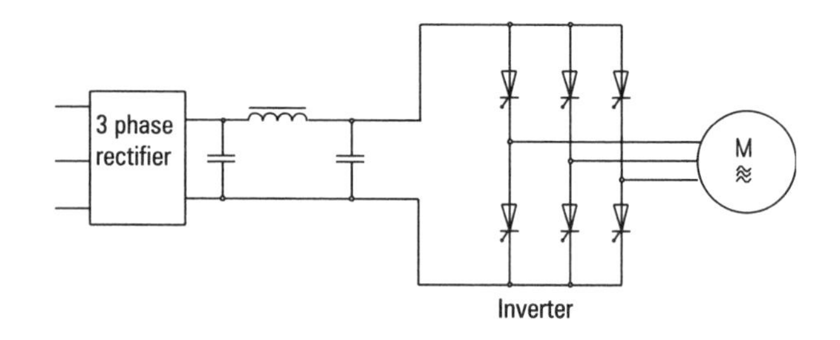

Figure 7.67 *3 ph. induction motor speed control*

The triac is a bi-directional switching device.

Its main advantage is that it can be used in a.c. circuits as it switched on both halves of the supply cycle.

It is a relatively low power device but is widely used in such circuits as the lamp dimmer and as a variable speed control for hand-held tools.

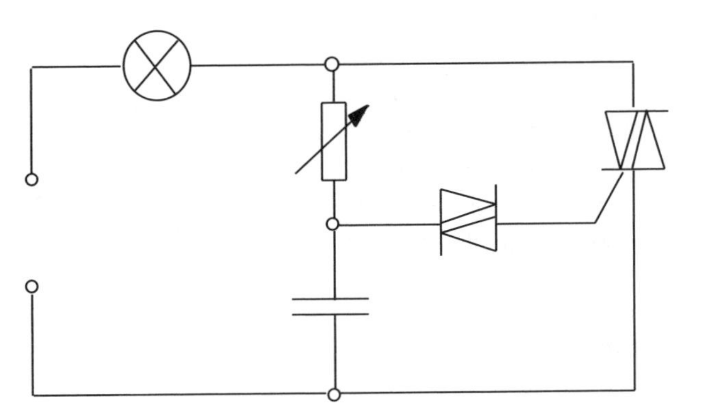

Figure 7.68

In the lamp dimmer circuit (Figure 7.68), the triac will fire as soon as the diac admits a pulse to the gate terminal. In "Lighting Systems", another book in this series, you will find the dimming of incandescent lamps.

The moment at which the gate pulse is admitted will then depend on the time taken for the voltage across the capacitor to reach the threshold value of more than 30 V. This in itself will depend on the time constant of the R/C circuit which is controlled by the setting of the variable resistor.

Some precautions must be taken when using thyristors or triacs in the switching of inductive loads such as motors or discharge lighting.

On breaking the load circuit, an induced back e.m.f. will be developed and unless this is suppressed, false triggering of the device may occur. This may be suppressed by means of a "snubber circuit" which is a simple R–C filter connected across the device (Figure 7.69).

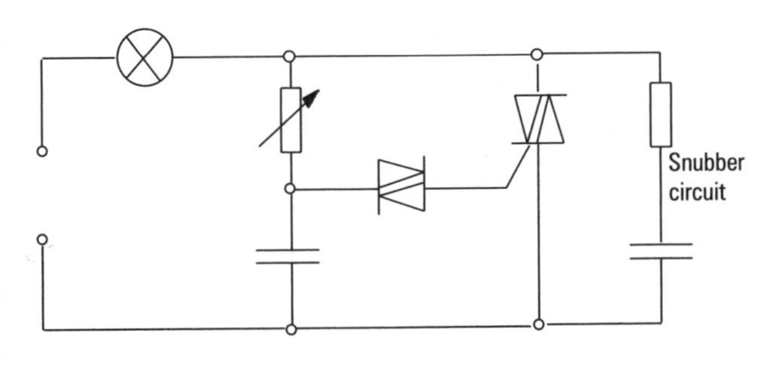

Figure 7.69

In the control of motors such as the series motor shown, a diode may be used. This is known as a "flywheel diode". The induced e.m.f. is suppressed by being fed back into the load.

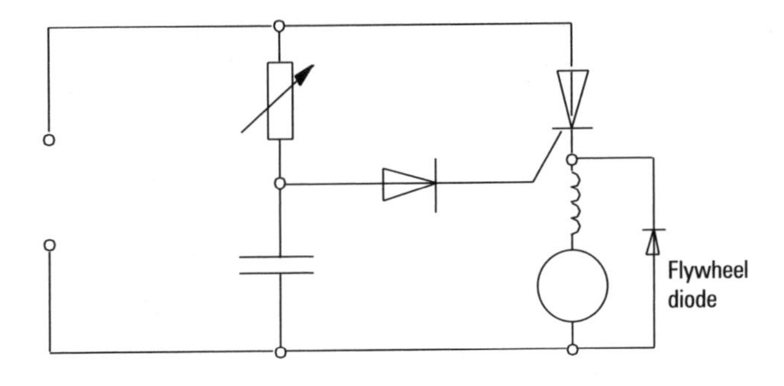

Figure 7.70

Testing of electronic and semiconductor devices

In the testing of electronic components, great care must be taken to ensure that the devices are not harmed by the test process.

Pay strict attention to the manufacturer's specification in terms of voltage, current and polarity. Take care also to ensure that the temperature of the device is not exceeded, as the leakage current in semiconductor devices will increase with temperature and a thermal runaway situation can quickly develop. Where heat sinks have been used, these should be left in place to dissipate any heat generated.

By using the resistance range of a multimeter, simple devices such as diodes and transistors can be tested for open or short circuits.

Diode testing

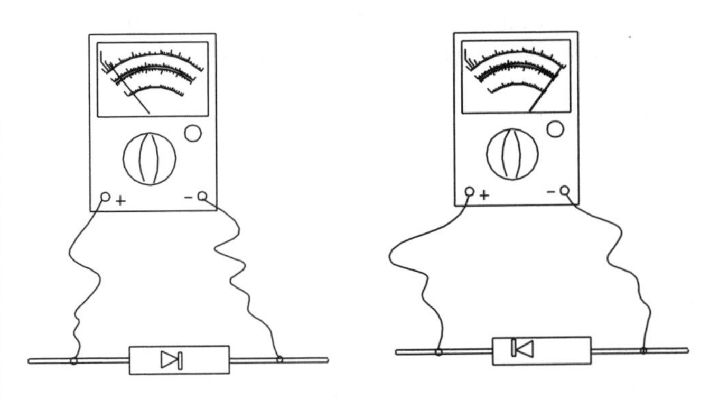

Figure 7.71 Testing diodes

Transistor testing

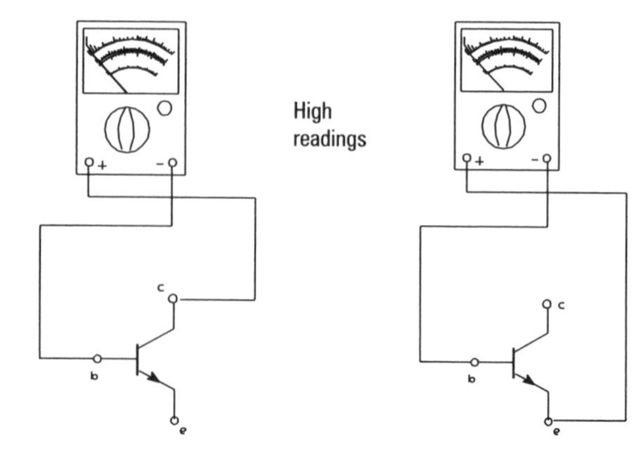

Figure 7.72 *Testing transistors*

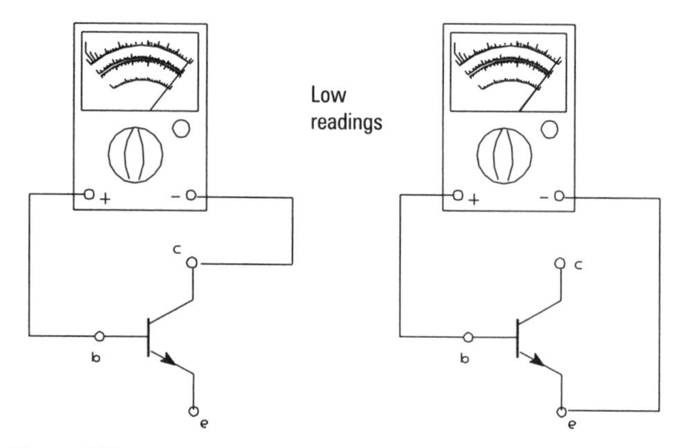

Figure 7.73

The polarity of some multimeters on the resistance ranges may not be as indicated by the terminals of the instrument. A clear indication of this will be when all your diodes and transistors appear to be "the wrong way round" when tested. If this happens, reverse the leads and carry on as normal.

Testing a thyristor or triac control on a motor circuit

It is not easy to test the control device by means of a multimeter. If the motor is not working, short out the control device with a temporary wire link. If the motor runs at full speed when switched on, the circuit is at fault.

If the motor does not run then it may be at fault.

Fault finding charts

The equipment manufacturer may provide the user with an easy-to-follow fault finding chart which can be used to detect and locate certain common faults in the equipment.

Before testing, a visual check should be made so that any signs of trouble may be identified:

Check that;

- all components are in place
- there are no signs of burning or overheating
- there are no loose wires or leads
- there are no obvious breaks in the circuit
- there are no accidental shorts
- all soldered joints are sound.

Testing

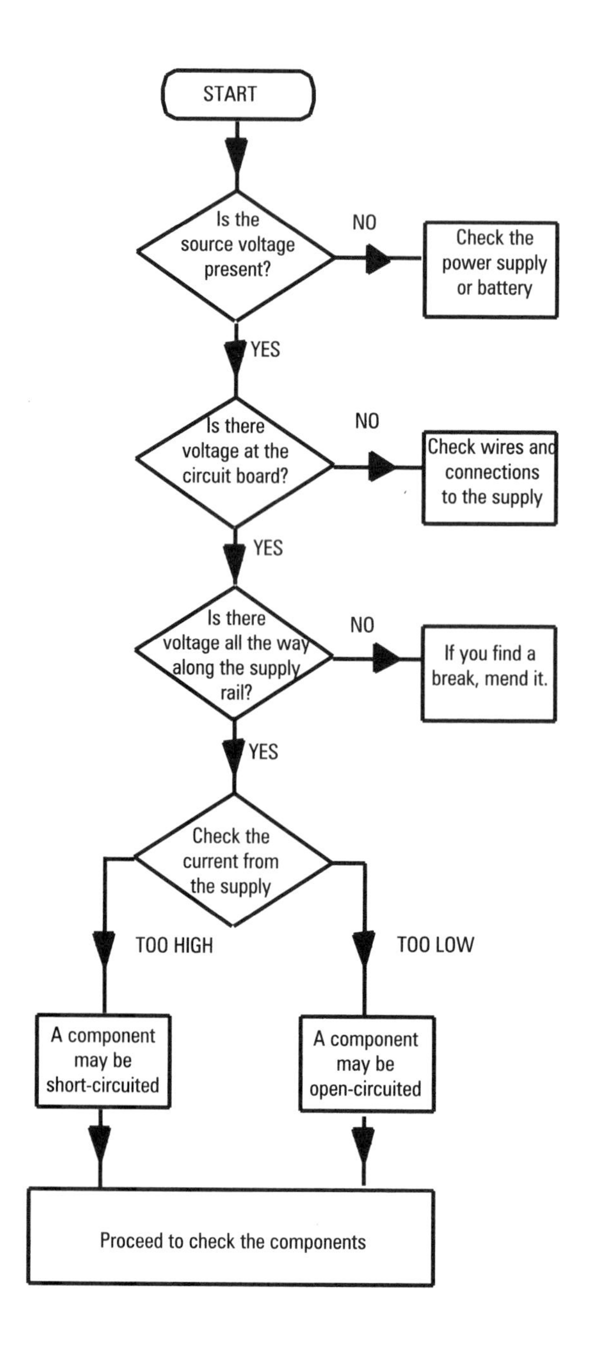

Component checks

A multi meter can be used for this purpose but remember

- in using it as a voltmeter you may be introducing a current path through the meter) where none existed previously. Also when connected across a high value resistor your instrument affects its value. Only use a voltmeter with a high ohms per volt ratio.
- in using it as an ohmmeter you are introducing voltage into the circuit and this may affect the behaviour of the components. Disconnect components before before measuring their resistance.

In signal and pulse mode circuits use an oscilloscope. This conveys far more information about the circuit such as amplitude, frequency, pulse shape and mark/space ratio.

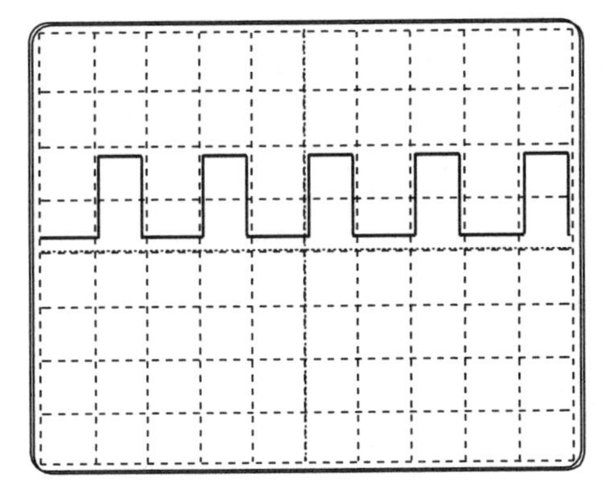

Figure 7.74

The simple go/no-go transistor test as previously described, may not be sufficient at for proper diagnostic testing and a commercial tester may have to be used. This will mean that the transistor will have to be removed from the circuit but the test result will be far more comprehensive.

More complex devices such as integrated circuits are normally tested by replacement with another similar device. This should only be carried out if the circuit conditions are correct and in accordance with the manufacturer's specification otherwise the replacement device may well end up in the same damaged state as the original.

Safety and security equipment

In the installation of safety and security equipment several types of remote sensing device are to be found.

Light beam

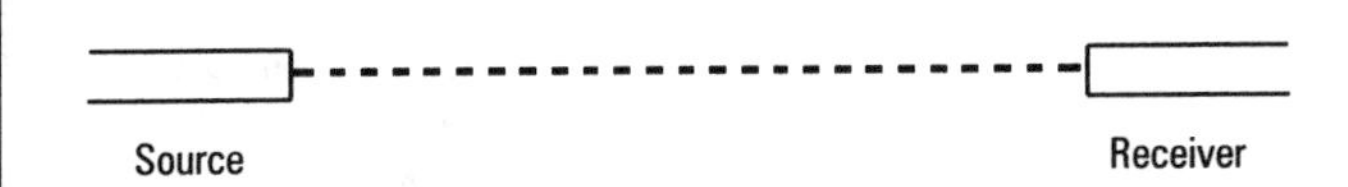

Figure 7.75 *Direct*

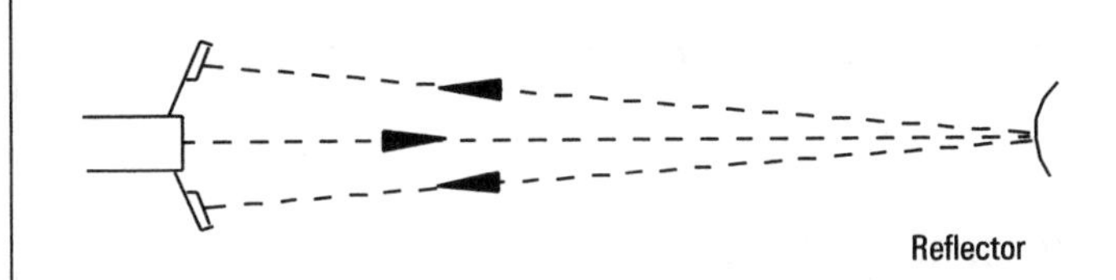

Figure 7.76 *Reflected*

A light beam falls onto a light-sensitive device which indicates that the condition is "normal". A break in the light beam will trigger a signal which will then initiate the appropriate response.

Infra-red beam

This device uses part of the light spectrum which is not visible to the naked eye and as a result finds application in security systems. The transmitter unit emits a pulse-modulated beam to a parabolic receiver unit. Any disruption of the signal will bring about a response in the receiver circuit.

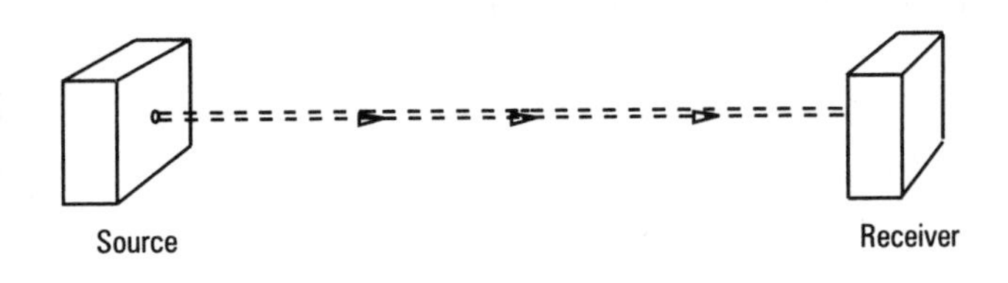

Figure 7.77 *Infra-red beam*

Passive infra-red

This type of detector is most commonly found on domestic "security light" fittings.

Despite its low cost, this is a very sophisticated form of heat detector which detects the heat radiated by any body which comes within its operating range.

The heat thus detected is focused by an arrangement of mirrors on to a thermo-electric sensor which then transmits a signal to the switching circuit.

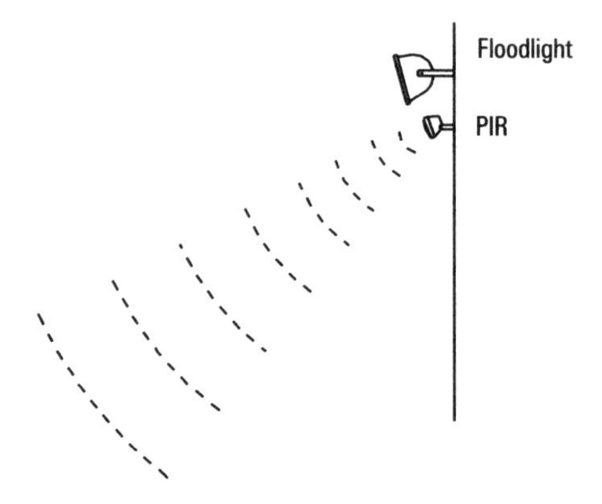

Figure 7.78 *Passive infra-red*

Ultrasonic detectors

These are used in applications where heat, light or microwave transmissions are unsuitable or unreliable. The transmitter emits pulses of a high frequency sound wave which is well beyond the range of human hearing. When these pulses strike a solid object, an echo returns to a receiver. The time delay between signal and echo is converted into a digital signal which can be used for further control functions.

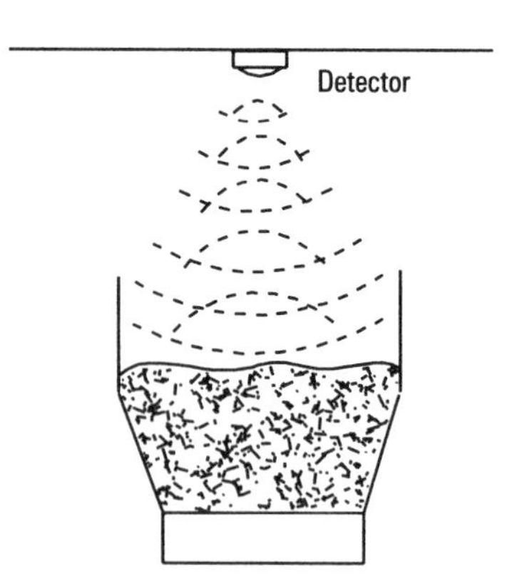

Figure 7.79 *Ultrasonic detector*

Exercises

1. It is sometimes necessary to supply equipment with a "stabilised supply".
   (a) State what is meant by the term "stabilised supply".
   (b) Draw a labelled circuit diagram showing how a 24 V stabilised d.c. supply may be obtained from a 230 V, 50 Hz input.
   (c) Explain the operation of the circuit in (b).

2. (a) State the precautions which should be taken when
   i) removing and replacing a printed circuit board from equipment.
   ii) removing and replacing components on a printed circuit board
   iii) testing electronic components.
   (b) Explain the difference between
   i) an LDR and an LCD
   ii) a thyristor and a thermistor
   iii) a transistor and a triac

3. (a) i) Draw the circuit symbol for a 2 input OR gate and a 2 input NOR gate
   ii) Construct truth tables for these two gates.
   iii) State the requirements for each of these gates to give an output.
   (b) Give practical examples of switches used as gates for
   i) AND
   ii) OR

4. (a) Describe the operational function of a
   i) thyristor (SCR)
   ii) junction transistor
   (b) Draw the circuit symbol for EACH device in (a) and label all connections.
   (c) Explain the use of a diac in a triac circuit.

End Questions

1. A timing circuit consists of a 1 MΩ resistor in series with a 20 μF capacitor connected across a 120 V d.c. supply.
   (a) Calculate the time constant of the circuit.
   (b) Construct a graph showing the rise of voltage across the capacitor with respect to time.
   (c) Use the graph to estimate the time required for the voltage across the capacitor to rise to 90 V.
   (d) Determine the energy stored when the voltage across the capacitor is 90 V.

2. (a) With the aid of a fully labelled diagram explain how the power factor of a large three-phase motor may be improved by directly connected capacitors.
   (b) A 50 Hz single phase inductive load is to have a parallel capacitor connected across it to correct the power factor. Instruments connected to the circuit show that the current is 53 A, voltage 230 V and a wattmeter shows 6360 W. Calculate the value of the capacitor to improve the power factor to 0.95.

3. A balanced, three phase star connected load draws 32 A from a 50 Hz 440 V supply at a power factor of 0.65 lagging. Assuming the load to consist of resistance and inductance only calculate
   (a) the impedance
   (b) the resistance
   (c) the inductance
   in each phase.

4. (a) Describe, with the aid of sketches, a current transformer in which the primary winding is a straight bar conductor.
   (b) It is required to measure the power taken by an 1800 kVA, 11 kV, three-phase, 3 wire load.
   Draw a circuit diagram to show how the power can be measured using two wattmeters, each rated at 5 A and 110 V.

5. (a) Draw up truth tables showing the difference between a 2 input AND gate and a 2 input OR gate.
   (b) The internal circuit of a washing machine requires that before a motor can operate the following conditions must exist:
   i) a programme must be selected (programme switch ON)
   ii) door must be in closed position (door switch CLOSED)
   iii) water level must be sufficient (water pressure switch OPEN)
   Draw a block diagram to show how logic gates may be used to control the washing machine motor.

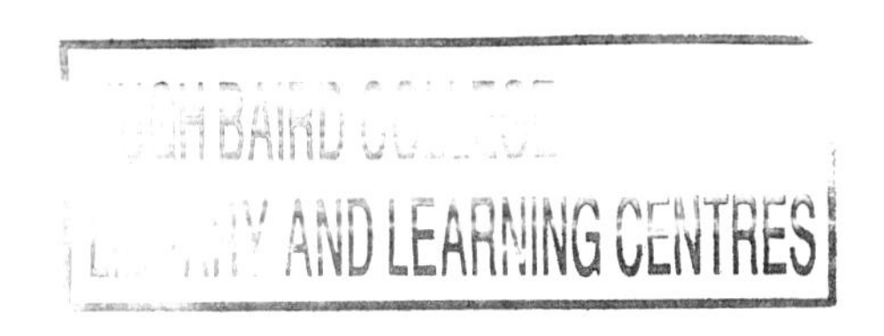

Answers

The descriptive answers are given for guidance and are not necessarily the only possible solutions.

Chapter 1

p.4 Try this: (1) 3.02 T, (2) 0.25 mWb, (3) 4.7 cm
p.5 Try this: (1) 375 mWB, (2) 25.57 Wb
p.7 Try this: 0.3316 A
p.13 Exercises: (1)(b) B – Teslas, H – Ampere turns per metre, (2) (b)i). 0.1 s, ii) 2 A, iii) 1.6 A approx., (3)(a) 0.101 A

Chapter 2

p.15 (1) 1.214 T, (2) μ_0, (3) direction of motion, (4) time constant
p.17 Try this: 24 mC
p.17 Try this: 72 μF
p.19 Try this: (1) 48 μF, (2) 12 μF
p.20 Try this: 32 Joules
p.23 Try this: 5.2 kV
p.24 Try this: (a) 650 V, (b) 5 secs
p.25 Try this: (a) 8 V, (b) 210 secs
p.27 Exercises:(1)(a) 5.714 μF, (b) 0.457 J, (c) 10 μf = 228.56 V, 20 μf = 114.28 V, 40 μf = 57.14 V, (d) 5.714 sec,
p.28 (2)(a) 0.8 secs, (c) approx. 0.8 secs, (d) 20 μf = 40 V, 80 μf = 10 V

Chapter 3

p.29 (1) When calculating the combined value – parallel capacitors behave in a similar manner to series resistors and series capacitors behave in a similar manner to parallel resistors.
(2) dielectric (3) to discharge the capacitor safely
p.32 Try this: (1)(a) 120 Ω, (b) 600 Ω, (c) 180 Ω, (2)(a) 30 mA (b) 1800 Ω
p.36 Try this: (a) 8.863 A, (b) 2.454 A, (c) 6.4 A, (d) 38.45 A
p.37 Try this: A–B = 114.58 A, B–C = –10.42 A, C–A = –160.42 A
p.38 Try this: (a) A–B = 91.25 A, B–C = 31.25 A, C–D = –18.75 A, D–A = –58.75 A, (b) 7.995 V, (c) 101.56 W
p.41 Try this: 1493.8 m
p.42 Try this: R_x = 225 Ω
p.45 Try this: (a) X_L 400 Ω, (b) X_C 100 Ω, (c) Z 500 Ω
p.46 Try this: (a) 20 Ω, (b) 31.83 Ω, (c) 15.5 Ω, (d) 3.225 A, (e) 104 W, (f) 0.645, (g) 49.8°, (h) U_R = 32.25 V, U_L = 64.5 V, U_C = 102.65 V
p.48 Try this: 450 A
p.52 Try this: (a) 2.309 A, (b) 6.93 A
p.53 Try this: (1) unity power factor, $Z = R$, current maximum, $U_L = U_C$, (2) 71.17 Hz
p.56 Try this: I_R = 10 A, I_L = 1.157 A, I_C = 4.146 A, I_T = 10.44 A
p.58 Try this: (a) 3036 VAr, (b) 36.87°
p.59 Exercises: (1) 27.27 m, (2) 63 A, (3) 36.4 mA, (4)(a) 8A, 3683 W, 0.697 p.f., (b) 24.13 A, 11652 W, 0.697 p.f.

Chapter 4

p.61 (1) R_2 = 120 Ω, V_2 = 30 V, (3) 72.17 A
p.67 Try this: (a) 20 Ω, (b) 4 A, (c) 0.8, (d) 32 Ω, (e) 2.5 A
p.68 Try this: 12.5 kVAr
p.70 (a) 360 μF, (b) 120 μF
p.71 Try this: 203.7 μF
p.72 Try this: Approx. 42 kVAr
p.75 Exercises: (1)(b)i) 53 kVAr per phase, ii) 1054.4 μF, (2)(b) 5.9 μF
p.76 (3)(b) 145.23 μF, (4)(c) 7.32 kVAr per phase

Chapter 5

p.77 (1) supplying any additional power, resources, economics (2) 4.8 kVAr (3) 8.94 kVAr

p.78 (a) $N_1 = \frac{I_2 \times N_2}{I_1}$ (b) $I_2 = \frac{U_1 \times I_1}{U_2}$ (c) $N_2 = \frac{N_1 \times U_2}{U_1}$

(d) $I_1 = \frac{I_2 \times N_2}{N_1}$

p.80 Try this: 39.2 A
p.84 Try this: (a) 75%, (b) 75%, (c) 0.707
p.85 Try this: (1) 2.1%, (2) 415 V
p.86 Try this: 8.8 kA
p.87 Try this: 1315 A
p.89 Try this: 400 V
p.92 Exercises: (1) (b) 4.424 kW, (2)(b)i) 95.6%, ii) 94.78%, (3)(b) 1.19%, (4) (b) 19.32 kVAr

Chapter 6

p.93 (1) 7.65 A (2) iron losses = copper losses (3) the autotransformer has one winding, the double wound transformer has two.
p.96 Try this: 0.0008 Ω
p.97 Try this: (a) 0.0020004 Ω, (b) 0.0010001 Ω, (c) 0.0006667 Ω
p.98 Try this: 399990 Ω
p.99 Try this: (a) 999950 Ω, (b) 1999950 Ω
p.101 Try this: 78.2 Ω
p.103 Try this: (a) 216 W, (b) 0.864 W
p.105 Try this: 250 Hz
p.107 Try this: 37.5 VA

p.108 Try this: 209.1 V
p.108 Try this: (a) 50000 Ω, (b) 0.3 mA
p.112 Try this: 75 Ω
p.114 Try this: 1600 W, 0.654
p.114 Exercises: (1)(c)i) 2600 W, 0.62 p.f., ii) 1900 W, 0.388 p.f., (3)(a) 6 Ω, (b) 0.1 A, (c) 0.3 A

Chapter 7

p.117 (1) safety, practicality, economy, (2) 1.1, (3) 0.92
p.119 Try this: (a) 10.6 V, (b) 119 Hz
p.129 Try this: 8 Ω

End questions

p.140 (1)(c) 25 s, (d) 81 mJ, (2) (b) 500 μF, (3)(a) 7.9375 Ω, (b) 5.16 Ω, (c) 19.1 mH